Praise for *New Space Capitalism*

"Market forces are revolutionizing space and the space economy, and Rainer Zitelmann's sweeping new book stands out from the crowd in unapologetically making the case for keeping capitalism at the core of our spacefaring future. A must-read for anyone interested in the economics of space and humanity's exploration of it."

—Matthew Weinzierl, professor of business administration and space expert at Harvard Business School

"Countering the pessimism that afflicts many analysts today, Dr. Rainer Zitelmann shines a bold and optimistic light on an emerging industry that promises to transform the future of mankind. His *New Space Capitalism* explores how private entrepreneurial activity is emerging as the force that will take humanity back to the Moon, to Mars, and beyond. He shows how private property rights give the incentive for space exploration, tourism, and industrial development and space settlement. This is a thrilling and very timely look at how capitalism can take humanity into its next great leap."

—Madsen Pirie, president of the Adam Smith Institute, London

"Rainer Zitelmann's book *New Space Capitalism* is one of the most recent, compelling, and best illustrations of how the limitless expansion in the future of humankind depends on the final triumph of the anarcho-capitalist system, based on entrepreneurial creativity unblocked by bureaucratic apparatuses and the ideology of statism. In Zitelmann's book, full of analytical insights from the Austrian School of Economics, you will find why space is not only 'the last frontier' of humanity, but also of human liberty."

—Jesús Huerta de Soto, professor of political economy, King Juan Carlos University, Madrid

"A fantastic book! It tells us compellingly that space exploration is an entrepreneurial activity and thus can be done best by entrepreneurs like Elon Musk, instead of government agencies. Space capitalism is more cost-effective and innovative than space statism. It is the only way humanity will ever be able to colonize Mars or mine asteroids. What the space sector needs most is full-scale decisive deregulation and not more regulations."

—Weiying Zhang, professor of economics, Peking University

"This book reports on the next major topic shaping the future of our economy. Using the example of private spaceflight, Zitelmann convincingly shows why entrepreneurship is superior to the often greatly overrated state-led industrial policy."

—Professor Hans-Werner Sinn, former president of the ifo Institute, Munich

"The challenge of opening the space frontier can only be met by the full power of human creativity. Freedom will give us space. Space will give us freedom. Zitelmann is showing us the way."

—Dr. Robert Zubrin, president of the Mars Society

NEW SPACE CAPITALISM

NEW SPACE CAPITALISM

THE ENTREPRENEURIAL PATH TO THE STARS

RAINER ZITELMANN

FOREWORD BY NEWT GINGRICH

Skyhorse Publishing

Skyhorse Publishing books may be purchased in bulk at special discounts for sales promotion, corporate gifts, fund-raising, or educational purposes. Special editions can also be created to specifications. For details, contact the Special Sales Department, Skyhorse Publishing, 307 Fifth Avenue, 4th Floor, New York, NY 10016 or info@skyhorsepublishing.com.

Visit our website at www.skyhorsepublishing.com.
Please follow our publisher Tony Lyons on Instagram @tonylyonsisuncertain.

10 9 8 7 6 5 4 3 2 1

Library of Congress Cataloging-in-Publication Data is available on file.

Cover design by David Ter-Avanesyan
Cover photo: Getty Images

Print ISBN: 978-1-5107-8821-3
Ebook ISBN: 978-1-5107-8822-0

Printed in the United States of America

CONTENTS

FOREWORD

BY FORMER SPEAKER NEWT GINGRICH

New Space Capitalism Really Matters

Rainer Zitelmann has made a major contribution toward thinking about the future of humans in space. His new book, *New Space Capitalism*, is a very important and comprehensive look at the next great breakthrough we need if the momentum of human development in space is going to reach its fullest potential.

We have had a tragic loss of momentum from the time Americans landed on the Moon to the present efforts to return to the Moon. A major reason for that loss of momentum was the degree to which space exploration became a function of governments, politicians, and bureaucracy. The result was a number of limited ventures, many of them technologically very elegant, but no accelerating momentum of human and financial energy breaking loose and creating a wave of exploration, occupation, and development.

It is as though the original cavalry expeditions that mapped the American West had not found any wagon trains to follow up and settle the areas they were exploring. Of course, in our history there were over 30,000 maps of the new regions published and given away to hopeful settlers.

Elon Musk's development of the reusable rocket and the dramatic collapse in launch costs (up to 90 percent and that probably understates it because, as a monopoly, SpaceX has a higher price than it will have when other reusable rockets are competing with it) has begun to transform the entire potential of space. As I write, the Starship is about to have its twelfth test launch. When Musk finally gets it to work on a regular basis it will revolutionize our concept of space. One general at Space Command compared Starship to launching a C-17 cargo plane into space on a regular basis. With either one hundred passengers or one hundred tons of supplies, this will be a revolutionary moment. Other companies are working overtime to match or exceed SpaceX's achievement.

Zitelmann understands that these developments create the possibility for an entire new economy in space, and that new economy will require a new legal structure.

I suggested this reality as a second-term congressman in 1981 when I introduced the Northwest Ordinance of 1787 as a structure for statehood for the Moon. That act provided that when a territory had 20,000 inhabitants it could organize and then when it had the same number of citizens as our smallest state (today that would be Wyoming with 590,000) the Moon could apply to become a state. I knew at the time it would be ridiculed but I wanted to express optimism for humans in space and suggest that we should begin to think about the legalities of their existence.

The important thing about Zitelmann's *New Space Capitalism* is that it really explains the fundamental requirement for the rule of law and some system of structures to enable private property and free enterprise capitalism to operate.

People are not going to invest if they think their money can be capriciously taken by the local boss or by a dishonest business.

Having a stable and reliable system of the rule of law is the absolute sine qua non for substantial investment.

Zitelmann has really opened the door to a much more profound revolution in how we think about and invest in space. Up to now the rarity and expense of rockets and the almost total reliance on government has led to policies which missed the great drivers of profit, income, and long-term wealth creation.

The economic passions which had led to the opening of the West simply were not part of the space psychological ecosystem from Apollo to just a few years ago. The development of reusable rockets and the plans for literally daily launches of Starship and other rockets is beginning to change everything. Suddenly space is on the verge of moving from bureaucrats and space specialists into an adventure involving entrepreneurs, marketing agents, customers, investors, and all the varied aspects of free enterprise capitalism.

Once space begins to enter the commercial-entrepreneurial mindset, things will happen very quickly. The development of the nuclear engine will make travel to Mars and beyond and mining of the asteroids vastly easier.

The United States should adopt a policy of 100 percent write-off for manufacturing and tourism in space and no taxes for forty years on profits from manufacturing and space tourism. Given these double tax advantages there will be a flood of new investors—the equivalent of the wildcatters who in the first half of the twentieth century made the oil and gas industry so dynamic.

There are over 900 billionaires in the United States today. Their investable wealth dwarfs anything NASA ever dreamed of. Given the right tax and regulatory environment, a number of them will follow Jared Isaacman in traveling into space and investing in space.

Zitelmann has outlined many of the challenges and opportunities and *New Space Capitalism* is an important contribution to where we need to go to develop a better future.

February 4, 2026

INTRODUCTION

On July 20, 1969, the day the first men landed on the Moon, and just a few weeks after my twelfth birthday, I published a special edition of my school newspaper, a publication about space travel, rockets, and astronomy called *Galaktische Zeitung*, which I sold to my classmates. The front page of this celebratory issue featured the U.S. flag and the headline: "3 Men Aiming for the Moon." The article stated: "Today, for the first time ever, man is going to set foot on the surface of the moon." Inside, I recommended a book about the Moon, which noted: "The world in which we live, our Earth, is beautiful, but it is too small for the human spirit."

My kid's room was filled with piles of astronomy books. An oversized map of the Moon and a poster of the Saturn V rocket that was bigger than I was hung on the walls. While other children dressed up as cowboys or Native Americans for carnival, I put on my space helmet and the silver space suit my mother had made for me.

Back then, I not only devoured astronomy books, I was also an avid reader of science fiction novels by Hans Dominik, often referred to as the "German Jules Verne." I was fascinated by his tales of ingenious engineers who transformed the world with their courage, ingenuity, and technical prowess. In Dominik's eyes, engineers were more than technicians; they were the shapers of the future, embodying the belief in human spirit, achievement, discipline, and the power to create greatness through their knowledge and hard work.

My passion for space exploration was ignited at the age of nine with the premiere of a popular science fiction TV series in 1966. I can still recite the prologue, which began: “What may sound like a fairy tale today might be tomorrow’s reality.” The series depicted robots mining raw materials on asteroids and humans settling on other celestial bodies. For me and my friends in the “Astronaut Club” I had founded, this vision didn’t sound at all like a distant fantasy but a future we would all experience at the beginning of the next millennium.

Following the Moon landing, there was widespread belief that mankind would soon be going to Mars. The conquest of space had begun. In fact, just two weeks after the Apollo 11 mission returned from the Moon, rocket pioneer Wernher von Braun, whose Saturn V rocket had taken humans to the Moon, presented NASA’s Space Task Group with a detailed plan for a manned mission to Mars, targeting a launch in 1981. The head of NASA at the time, Thomas Paine, even specified an exact date: Twelve astronauts were to embark on a mission to Mars on November 12, 1981.[1]

But things turned out differently. After only five more Moon landings, the program was aborted—and Mars has yet to be reached. So, what happened? In his book *Spacefarers*, Christopher Wanjek writes: “I and others would argue that the primary reason NASA has done too little in *human* space exploration for the last forty years is that the space agency is directed by ever-changing US presidents (twelve since its creation) and micromanaged by the US Congress. The ‘vision’—stay in low-earth orbit; no, go to the Moon; no, go to Mars instead; no, go to the Moon—is a blurry, moving target. Jimmy Carter pushed for space science over human activities;

Ronald Reagan supported the ISS as a stepping-stone to a larger presence in orbit; George H. W. Bush pushed for a return to the Moon and then a journey to Mars; Bill Clinton focused on cooperation with the Russians to complete the ISS, which was over budget and less international when he took office; George W. Bush wanted to return to the Moon and Barack Obama wanted to skip the Moon and go to asteroids and Mars. Donald Trump has advocated, at various times, to go to the Moon, or to Mars, and to create a Pentagon-led space force."[2]

And most recently, Donald Trump promised that humans would land on Mars during his second term in office.[3]

I would respectfully disagree with Wanjek on the details here—Obama's only virtue was that he had zero interest in space travel and was therefore more than happy for private companies like SpaceX to take the lead. His disinterest was what really paved the way for private space exploration. However, Wanjek is undoubtedly correct in asserting that political interference stymied progress in manned space exploration. This is evident from NASA's unmanned space program, where reduced political involvement has allowed scientists to chart a more successful course. But the underlying reasons for the failure of government-led space exploration are more profound, as I will explore in this book.

The absence of clear goals was one of the reasons NASA became increasingly unambitious. Without a big, worthwhile goal, there is little incentive to take risks. In his book *Safe Is Not an Option*, space technology and policy expert Rand Simberg discusses how government-led space initiatives have become increasingly risk-averse, which, he claims, has paradoxically led to even greater risks in some instances. Are we, as a society, still willing to take risks? The ongoing debate about a mission to Mars or the colonization of Mars, as

I describe in chapter 5 of this book, certainly raises doubts on this score.

Simberg explains: “No frontier in history has ever been opened without risk and loss of human life and the space frontier is no different. . . . Historically, had we taken the same attitude toward safety in expanding our ecological range as we have in space, humanity would never have left Africa for Europe and Asia, colonized the Arctic, opened up the Americas, Australia, the south Pacific Islands, or settled the American West. We would not have made great scientific discoveries, developed steam engines, or steam ships, or rail, or automobiles, or aircraft. In fact, we would still be sitting in the trees, gazing out at the savanna and wondering what it might have in for us.”[4]

Today, I am cautiously optimistic, largely due to entrepreneurs such as Elon Musk, who not only articulate ambitious goals in space exploration but have also made significant strides toward achieving them. They are willing to take risks because they are entrepreneurs and because someone like Musk is pursuing a truly incredible vision: the colonization of Mars. Within the next decade, the dream of landing on Mars could become a reality—not thanks to government space programs, but due to the efforts of private companies.

Following the Moon landing, government space programs stalled. For four decades, the cost of launching a kilogram of payload into space stagnated while Elon Musk, as we will see, reduced these costs by over 90 percent in just a few years. And that’s just the beginning. Due to his de facto monopoly, Musk currently passes on only a portion of these cost savings to his customers. As soon as he has more competitors (and at the same time continues to reduce production costs), prices will continue to decline for all market participants.

The following recent example highlights the rapid pace of development: In April 2025, the journal *Aerospace* published a study that included a comparative analysis of launch costs between reusable and traditional launch vehicles. The study considered various factors, including the number of times a rocket can be reused, assuming a reuse rate of ten times, "taking into account the current technological level."[5] By February 2026, a single Falcon 9 rocket had already been reused thirty-three times.

High launch costs were also a result of the industry's absurd "cost-plus contracts," which I will discuss in more detail later. These contracts, typically between NASA or another government space agency on the one hand, and a private manufacturer on the other, provided no incentive whatsoever for companies to reduce costs. Instead, they encouraged companies to drive costs up. Government officials, concerned about excessive corporate profits, required detailed cost documentation and allowed companies to add a profit margin of approximately 8 to 10 percent—whatever the begrudging government officials deemed appropriate.

The result is described by the space specialist Robert Zubrin as follows: "In the free enterprise world, manufacturers increase profit by cutting costs. In the cost-plus-contractor world, manufacturers increase profit by increasing costs."[6] The remarkable cost reductions achieved by SpaceX and others can be attributed, at least in part, to their adoption of fixed prices for services instead of cost-plus contracts.

The success story of private space exploration stands in stark contrast to the prevailing global trend over the past fifteen years. During this period, there has been a notable shift toward increased government intervention and less free market—a stark contrast to the 1980s and 1990s, when Margaret Thatcher, Ronald Reagan,

Deng Xiaoping, the Vietnamese "Đổi Mới" reforms, and the post-socialist transformations in Eastern Europe broadly promoted capitalism. Today, however, the "Index of Economic Freedom" is at almost its lowest level since it was first compiled in 1995.[7] In China, Xi Jinping is steering a course back toward greater government intervention, Latin America has experienced a resurgence of socialism, and even Chile, once the continent's shining economic beacon, was governed by a socialist. Only Argentina, under Javier Milei, is taking a different path. In Europe, there is also a growing trend of government intervention, signaling a move toward a centrally planned economy where politicians and bureaucrats, rather than businesses and consumers, call the shots.

In one area, however, developments are moving in exactly the opposite direction: space exploration. Here, the superiority of entrepreneurial initiative over bureaucratic apparatuses is evident as rarely before. Without SpaceX, American astronauts would still be dependent on Russian Soyuz spacecraft to take their astronauts to the International Space Station. And the prospect of reaching Mars would remain stubbornly out of reach.

For the Italian American economist Mariana Mazzucato, the Apollo program, with its Moon landings, is "an example of an extraordinary feat: government stepping outside the market-fixing role and into the market-shaping one."[8] Taking the Moon landing as an example, Mazzucato hopes to demonstrate that humankind's greatest achievements are accomplished when the state sets visionary goals and adopts an "entrepreneurial" approach to their pursuit. However, many scholars have since proved that this approach has failed far more often than it has succeeded.[9] This is particularly evident in the realm of manned spaceflight: Mazzucato picks out ten successful years (1961 to 1971) from the history of NASA, but ignores the failures of the following fifty years, including the space shuttle program, because it contradicts her thesis.

The space shuttle program began with an ambitious vision: the development of a reusable space shuttle that would drastically reduce the cost of traveling to space. However, the result was a rocket that cost more for each launch than conventional single-use rockets. Some experts at the time even concluded that the traditional concept of disposable rockets was perhaps more cost-effective after all. This conviction even prevailed until the early 2010s, when Elon Musk proved the opposite with his Falcon 9 rocket.[10]

Thus, the history of space exploration in no way supports Mazzucato's thesis of "value-makers and value-takers," which posits that governments are responsible for the creation of all essential products and private entrepreneurs merely steal them.[11] The example of Robert Goddard, the pioneer of American space travel, shows precisely the opposite—more on this in chapter 3. Later, Elon Musk developed the reusable rocket that NASA and other government space agencies had found such a tough nut to crack for decades. Musk developed his reusable rocket because he was convinced it made sense, not in response to any demands, instructions, or directives from NASA. Today, NASA is forced to buy services—such as the transport of goods and astronauts to the International Space Station—from Musk because it can't provide them itself.

Then there are the claims that SpaceX is entirely reliant on NASA, public subsidies, or government contracts—claims that are misleading. On August 21, 2025, Elon Musk stated on X that approximately $1.1 billion of SpaceX's revenues can be attributed to NASA, out of total sales amounting to $15.5 billion. So, in truth, NASA contracts represent only 7 percent of SpaceX's sales. Space expert Eugen Reichl estimates that other government contracts, such as those from the United States Space Force (USSF) and the National Reconnaissance Office (NRO), the U.S. agency responsible for spy satellites, provide an additional $1.5 billion. "If you add up all the government contracts, they account for 20 to at most 25 percent," estimates Reichl.[12]

The claim that SpaceX has "heavily benefited from federal and state support,"[13] is misleading. Firstly, more than three-quarters of SpaceX's revenues, potentially up to 80 percent, are derived from nongovernmental sources. Secondly, government agencies only awarded contracts to SpaceX because its services, typically, cost less than those of other providers. Previously, ULA, a joint venture between Boeing & Lockheed Martin, had almost always been the only potential supplier. Finally, it is important to distinguish between government *subsidies* on the one hand and *contracts* from government agencies and the military on the other, a distinction that many commentators fail to make.

One of the topics explored in this book is why, after the triumphant successes of the Apollo missions, government-run manned spaceflight endured decades of inertia and wasted money, and what entrepreneurs like Elon Musk have done differently. More importantly, however, it also examines the legal and economic frameworks necessary to transform visionary concepts into reality. My thesis: Space capitalism is the only way humanity will ever be able to colonize Mars or mine asteroids. Only when it is possible to own, buy, and sell land on other celestial bodies can humanity truly become a multi-planetary species. And without strong economic incentives, this is a goal we will never achieve. Chapter 10, which elaborates on this thesis, is the most important in this book, though to fully understand it, you will also need to read the preceding chapters.

In 1950, pollsters at the Allensbach Institute asked a representative sample of West Germans whether they believed humans would be able to visit other celestial bodies, such as the Moon, within the next fifty years, so by the year 2000. At that time, only one in four thought this would be possible. Even in a subsequent survey in 1965, just four years before the Moon landing, only a slight majority of 53 percent of respondents believed it would be possible to journey to another celestial body.[14]

Dreams can sometimes seem distant, but occasionally they are much closer than we think. On October 7, 1903, Samuel P. Langley attempted to launch his Great Aerodrome aircraft from a catapult ramp on a boat on the Potomac River. The aircraft immediately tilted after takeoff, its nose dipped sharply, and it crashed into the water, sustaining significant damage. In a derisive editorial on October 9, *The New York Times* referred to the failed attempt at flight as a "ridiculous fiasco" and made a bleak prediction: "[It] might be assumed that the flying machine which will really fly might be evolved by the combined and continuous efforts of mathematicians and mechanicians in from one million to ten million years."[15]

Just two months later, on December 17, 1903, the Wright brothers achieved the first successful powered flight in the town of Kitty Hawk, North Carolina. Initially, *The New York Times* did not report on this milestone; it was not until 1906 that the newspaper published a detailed article about the Wright brothers' accomplishment.

As private individuals, the Wrights had less than $1,000 at their disposal—equivalent to approximately $38,000 today—while university professor Langley had $73,000, or about $2.8 million in today's terms, with more than two-thirds of this funding provided by the U.S. Army.[16]

The Wrights were fortunate that there was no government agency at the time to impose strict rules and regulations on exactly how they should build and operate their aircraft. George Nield, the man responsible for regulating commercial space exploration at the U.S. Federal Aviation Administration (FAA) for many years, said in an interview: "I think the difference is that the government didn't design, build, own, and operate the Wright brothers' aircraft. Think about how aviation would've developed if that been the scenario! That's really what we were faced with for space in the early years."[17]

What the space sector needs most is full-scale decisive deregulation and not more regulations. Throughout this book, I will critically examine the arguments put forward by statists, who see things quite differently. In his 2022 book *Space Ethics*, Brian Patrick Green, one of many "space ethics experts," insists: "However, commercial and industrial access to space . . . should not be a free for all. Instead, these private organizations still need to be regulated and still need state guidance. . . . It is the duty of the public sector—governments, states, and international bodies like the UN—to regulate the private sector for the sake of defending the common good for all on Earth and in space."[18]

There is nothing fundamentally wrong with regulation, but unfortunately many of those who vehemently advocate it do not have the creation of a clear legal framework for competition in mind: They want something completely different. James S. J. Schwartz, another prominent "space ethicist," outlines his vision for a potential regulatory mechanism to govern mining in space. He calls for an international body—a scientific panel—to be convened by the UN to oversee access to celestial bodies and areas thereof. Proposals for mining, he argues, would first have to be submitted to the panel for approval. If approval were granted and resources extracted, Schwartz explains, these resources would then have to be distributed fairly by the panel, or any revenues from the exploitation of those resources taxed.[19]

His second suggestion relates to the use of any extracted resources, namely in support of what he calls ethically defensible projects, and not ethically indefensible ones. As a thought experiment to justify his claim, Schwartz states: "suppose that an incredibly wealthy space entrepreneur wanted to sculpt a one-km-tall likeness of their person out of lunar water ice."[20] Even if that entrepreneur were to pay in order to contract the necessary exploitation, and this massive amount of money were distributed "equitably" to all states, Schwartz objects that there must also be a mechanism in place to stop the pilfering of such an incredibly rare resource for a frivolous purpose. Hence Schwartz's calls for committees and panels. Just as on Earth, statists believe that politicians and civil servants know better than millions of entrepreneurs and consumers.

When I read about the ideas "space ethics experts" come up with, I am filled with apprehension about the future of space exploration. We cannot afford to underestimate the influence of statist and egalitarian ideologies, which have already wrought havoc on Earth, largely because their critics dismiss them as absurd and mistakenly play down the potential damage they and their proponents can do.

This book not only chronicles the development of space exploration since the Moon landing, it also envisions the possibilities of the next fifty years—at least if the ideologues and statists do not gain the upper hand in this domain, as they have in so many others. Above all, I hope that this book will resonate with younger readers. My biggest piece of advice to young people is this: Stop studying business administration or law! And don't even think of studying gender studies or anything like that! That won't get you anywhere. Study a subject related to space. That's where the future lies. Elon Musk recognized this as long ago as 2003, when he stated: "I like to be involved in things that change the world. The Internet did, and space will probably be more responsible for changing the world than anything else."[21]

Today, space companies are open to hiring people with completely different backgrounds, provided they are passionate about space and possess the requisite skills. Keep in mind: "We are progressively becoming a space-based society. . . . Space, therefore, represents today's new frontier and the backbone of tomorrow's economy."[22]

When I had already finished writing this book, reports emerged that SpaceX was seriously considering an initial public offering (IPO). According to recent media reports, the company has been weighing a possible mid-June 2026 IPO at a valuation of around $1.75 trillion, potentially raising up to $75 billion. If realized, such an offering would be the largest IPO in history and would give the topic of the New Space Economy an entirely different weight in public perception and in capital markets. Whether such an IPO ultimately takes place, and under what exact terms, remains uncertain. Regardless of the outcome, the broader development of the space economy is clearly accelerating at an ever faster pace.

CHAPTER 1

GOVERNMENT FAILURE—THE END OF THE FUTURE

On November 14, 2011, at 4:14 a.m., a Russian Soyuz FG rocket took off from the Baikonur Cosmodrome in Kazakhstan. It was a historic launch site in the state that had become independent two decades earlier after the collapse of the Soviet Union. Fifty years earlier, it was from this site that the Soviet cosmonaut Yuri Gagarin embarked on his pioneering journey as the first human in space. At the time, Gagarin's mission was the latest in a series of blows to the United States, coming as it did after the shock caused by the Soviet Union's launch of the first artificial satellite, Sputnik 1, into space in 1957. The USSR had also succeeded in launching the first dog into space, thereby paving the way for human spaceflight. Sputnik 1 was the first artificial satellite in human history and marked the dawn of the space age—as well as igniting the space race between the United States and the Soviet Union. It was a race that the United States famously won with the first moon landing on July 20, 1969. However, by 2011, the United States was reliant on Russian capabilities, unable to reach the International Space Station (ISS) without Russian Soyuz rockets.

This Soyuz FG launch, initially scheduled for late September 2011, had been postponed twice following a third-stage misfire of another Soyuz rocket in late August. To ensure the safety of all involved, two unmanned test flights were conducted before proceeding with the mission.

It was a special flight for Daniel Christopher Burbank. The forty-nine-year-old from Connecticut had trained as a helicopter pilot in Florida for eighteen months and become a pilot. In the early 1990s, he had applied to the National Aeronautics and Space Administration (NASA) to become an astronaut, but his first two applications were rejected. His persistence paid off with his third application, which led to his acceptance in 1996 and subsequent two-year basic training at the Johnson Space Center. Four years later, he made his first flight on the space shuttle to the ISS. Then, in 2006, Burbank boarded the space shuttle and flew to the ISS a second time, where he conducted his first extra-vehicular mission in space.

But this new mission was different. Burbank was flying with two Russian colleagues, Anton Shkaplerov and Anatoly Ivanishin, and the mission was called Soyuz TMA-22. The destination was once again the ISS, but with the discontinuation of the space shuttle program on July 8, 2011, following two tragic accidents that claimed the lives of fourteen astronauts, the United States no longer possessed its own means of reaching the ISS. Efforts to develop a replacement for the space shuttle had not been successful.

The flight was expensive, with the United States paying $43 million per seat, equivalent to approximately $58 million today. In just a few months, the price per seat had increased from $28 million to $43 million, a staggering rise of 54 percent—and that was just the beginning.[23] By October 2020, the cost had escalated to $90.25 million, or about $110 million today.[24] In other words, monopoly prices, which, in the opinion of the Americans, were by no means justifiable.

Time magazine described the cramped conditions aboard the Soyuz space capsule as follows: "Going into orbit aboard a Soyuz spacecraft is a humbling experience. There's no snazzy space-shuttle cockpit with wraparound windshield and heads-up displays here, no proper couches from which a true commander can fly a true rocket ship. The Soyuz is more like a hamster ball with three small seats stuffed in its bottom into which three adults squeeze themselves shoulder to shoulder. The legroom is so limited, they must fly with their knees drawn up toward their chests—rocket jocks in fetal position. The commander—on his back on the floor in his grownup car seat—can't even reach the instrument panel without the help of a sort of conductor's baton with a rubber tip."[25]

The diameter of the Soyuz launch and landing module was designed to be as small as possible, measuring just 2.17 meters. In comparison, the American crew exploration vehicle Orion, also a capsule, had a diameter of five meters.[26]

The Americans were less humiliated by the cramped conditions than they were by their total reliance on Russian transportation to get to the ISS in the first place. *Time* magazine also stated: "The U.S., meantime, has to tolerate Russia and pay its extortionate ticket prices simply because that's the only lift around. When you're hitching a ride to spring break, you'd better be ready to pay for the gas and the Red Bull and put up with whatever's on the radio. The ISS (despite that International in the moniker) is almost entirely a NASA-built, U.S.-paid-for station, and the hard fact is, Moscow controls our access to it."[27]

Despite escalating international tensions, Russia and the United States continued to work together on the ISS, although the tone between the two nations had become more abrasive, including on the subject space exploration.

Then, on February 27, 2014, Russia attacked Ukraine. Armed soldiers wearing uniforms devoid of national symbols ("green men"), later identified as Russian special forces, occupied the regional parliament and other strategic points on the Crimean peninsula. Following a contentious referendum, Russia announced in March that it was annexing Crimea. The United States and other Western states responded, albeit half-heartedly, with sanctions against Russia.

"After analyzing the sanctions against our space industry," tweeted Russian Deputy Prime Minister Dmitry Rogozin on April 29, "I suggest the USA bring their astronauts to the ISS using a trampoline."[28] Two weeks later, Rogozin announced that Russia would not be flying any astronauts to the International Space Station (ISS) after 2020, in other words until at least four years before the anticipated end of the space station's operational life. Additionally, Russia would no longer sell RD-180 engines to the United States if they were going to be used on rockets carrying military payloads.

NASA responded calmly: "Russia is every bit as dependent on us to operate the space station as we are on them," said Greg Williams, one of the chief administrators in NASA's Human Exploration and Operations office, adding that the dustup even provided NASA a small benefit. "We got a new vocabulary word—we didn't know the Russian word for trampoline."[29]

In fact, U.S. astronauts have continued to fly on Russian Soyuz rockets to this day, despite ongoing tensions between the two countries. The most recent manned Soyuz mission involving a U.S.

astronaut was Soyuz MS-28, which launched from Baikonur on November 27, 2025, carrying a crew of two Russian cosmonauts and NASA astronaut Christopher Williams to the ISS.

In any case, since 2020, the Americans have not had to rely solely on Soyuz rockets to fly to the ISS. In fact, most American astronauts, along with an average of two Russian cosmonauts per year, have been transported by the private company SpaceX. We will look at this in more detail in the next chapter.

But how did the United States ever find itself in this position, unable to transport astronauts to the ISS between 2011 and 2020? To understand this, we first have to go back to the 1970s.

The landing of twelve American astronauts on the moon stands as one of humanity's greatest—if not the greatest—achievements. However, it came at a significant cost. Between 1960 and 1973, the United States spent $25.8 billion on the Apollo program, equivalent to approximately $257 billion when adjusted for inflation in 2020. Including the Gemini project and the robotic lunar program, which were instrumental to Apollo's success, the total expenditure reached $28 billion ($280 billion adjusted for inflation). Spending peaked in 1966, three years before the first moon landing. NASA spent a total of $49.4 billion during this period ($482 billion adjusted for inflation in 2020).[30] In 1966, NASA's budget represented 4.5 percent of the total U.S. budget; today it is less than half a percent.

NASA's success demonstrated the superiority of capitalism, but only indirectly: The United States was only able to invest so heavily in its space program because its capitalist economic system was far superior to the Soviet Union's socialist system. However, the U.S. space program was primarily government-run. The Saturn V rocket was developed by NASA, with its construction and design coordinated by several companies under the leadership of Wernher von Braun and his team at the Marshall Space Flight Center in Huntsville, Alabama. Private companies, including Boeing, North American Aviation, Douglas Aircraft Company, and Rocketdyne, built the various rocket stages and engines. Unlike today, when companies like SpaceX independently develop rockets and sell services to private or government clients, the private companies back then were predominantly the executive bodies of NASA, which set very strict specifications.

When the first humans landed on the moon on July 20, 1969, Republican Richard Nixon had been the U.S. president for only six months. It was his responsibility to determine the course of action following the moon landing.

Nixon reaped the benefits of the groundwork laid by his predecessors, Lyndon B. Johnson and John F. Kennedy. It was Kennedy, president of the United States from 1961 to 1963, who, in a speech on May 25, 1961, set the ambitious goal of putting an American on the moon before the end of the decade. Kennedy positioned manned space travel as a central element of the competition between the United States and the Soviet Union during the Cold War and gave the political green light for the Apollo program. Although he was initially skeptical about the high costs involved, Kennedy soon recognized the project's strategic and symbolic value.

Lyndon B. Johnson, who assumed the presidency following Kennedy's assassination in 1963 and served until 1969, played a pivotal role in shaping American space policy. As senate majority leader and later as vice president, he was a staunch supporter of the space program and, as head of the National Space Committee, coordinated the implementation of Kennedy's vision. Under his presidency, critical funding was made available for the Apollo program, and he garnered political backing at a time of growing social unrest. Johnson's unwavering support ensured that Kennedy's ambitious goal would actually be achieved.

However, during the moon landing celebrations, Nixon notably refrained from mentioning either Kennedy or Johnson by name,[31] a decision that drew criticism from prominent newspapers such as *The Washington Post* and *The New York Times*. A commentary in *The New York Times* bearing the headline "Nixoning the Moon" captured the mood perfectly.[32] In anticipatory obedience, NASA had prepared a statement for the president which created the impression that the Moon landing was Nixon's achievement. This was too much for Nixon's space advisor Frank Bormann (the first man to orbit the moon in December 1968). He advised Nixon not to use the NASA draft: "Look, Mr. President, you really don't have anything to do with Apollo 11. You're just the fortunate or unfortunate recipient of this mission. . . . If it fails, you'll get tarred with it, and if it succeeds you'll get *some* of the credit. But for you to say what NASA is suggesting—that in effect you were the father of the space program—is just plain wrong."[33]

Following the Moon landing, thoughts turned to the next phase of space exploration. Just two weeks after the historic event, Wernher von Braun, the mastermind behind the Apollo program, presented the Space Task Group with a detailed plan for a manned mission to Mars: "The 1981 manned Mars mission (1982 landing on Mars) is shown as an integral part of the total space program for the next two decades. A 1982 manned Mars landing is a logical focus for the programs of the next decade. Although the undertaking of the mission will be a great national challenge, it represents no greater challenge than the commitment made in 1961 to land a man on the moon."[34]

Shortly before the Moon landing, Nixon's vice president, Spiro T. Agnew, had demanded that the next step should be taking humans to Mars.[35]

But Nixon was unwilling to support a Mars mission. Nixon—perhaps even more so than other politicians—was famous for paying slavish attention to opinion polls and aligning his decisions with public sentiment. The polls painted a clear picture: 53 percent of Americans were against a mission to Mars and only 39 percent were in favor.[36] As *The New York Times* opined: "Any forced draft Martian analogue of the Apollo project would divert hundreds of billions of dollars that are more urgently required to meet the needs of men and women on earth."[37] Or as Mike Mansfield, Democrat and senate majority leader, declared: Any such venture should not take place "until problems here on earth are solved."[38]

Nixon was concerned that endorsing a spectacular space program would allow his critics to argue that his government was wasting funds that could be better spent fighting poverty or solving other social issues at home. This argument, while popular, is entirely wrong. As I demonstrate in my book *How Nations Escape Poverty*, in which I draw on a wealth of scientific studies, neither increased government spending nor development aid are effective tools to combat poverty.[39] Nonetheless, this concern significantly influenced Nixon's decision-making.[40]

First, however, a decision had to be made about which of the upcoming lunar missions should still take place as planned. Just months after the first Moon landing, in November 1969, humans landed on Earth's closest neighbor for a second time. Then, on April 11, 1970, NASA launched Apollo 13 with the goal of landing in the Moon's Fra Mauro highlands. Fatefully, two days after takeoff, on April 13, a serious explosion occurred in one of the two oxygen tanks in Apollo 13's service module. The explosion, caused by a short circuit, destroyed the oxygen tank and disabled many of the craft's power and life-support systems. The astronauts—James Lovell, Jack Swigert, and Fred Haise—were forced to abort the moon landing and use their undamaged lunar landing module, *Aquarius*, as a temporary lifeboat.

Over the next few days, the crew, together with Mission Control in Houston, worked tirelessly to secure power, water, oxygen, and make the necessary course corrections. Despite extreme cold and CO_2 problems, they managed to get the space capsule back to Earth. On April 17, 1970, Apollo 13 splashed down in the Pacific Ocean, all three astronauts alive and well. Maybe you've seen the riveting documentary *Houston, We Have a Problem*, which tells the story of this ill-fated mission?

There were more flights to the Moon, but following the near-catastrophe of Apollo 13, President Nixon grew concerned about potential mishaps before the November 7, 1972, election. He even unexpectedly postponed the launch of Apollo 17, which was originally scheduled to take place before the election, due to warnings from his advisors about the mission's heightened risks. Nixon was worried that if something went wrong, he would be held responsible, potentially jeopardizing his reelection prospects.[41] The flight finally took place on December 7, 1972, exactly one month after Nixon's triumphant election victory against George McGovern.

But a statement made by Nixon at the launch of Apollo 17 came as a shock to all space enthusiasts. He remarked, "This may be the last time in this century that men will walk on the Moon."[42] When his comment was relayed to the Apollo 17 crew orbiting the Moon, astronaut Harrison "Jack" Schmitt was furious and thought to himself: "That was the stupidest thing a President could have said. . . .

Why say that to all the young people in the world. . . . It was just pure loss of will."[43] Historian John M. Logsdon was equally critical: "By his space decisions, Nixon made sure that his forecast would become reality."[44] Indeed, Nixon canceled all three subsequent planned Apollo moon landings (Apollo 18, 19, and 20), and any ambition to target Mars as the next destination was also abandoned.

But what vision should they follow? While they didn't want to abandon manned spaceflight entirely, perhaps NASA should redirect its attention? "President Nixon and his advisers," as historian John M. Logsdon explains in his book *After Apollo?: Richard Nixon and the American Space Program*, "were interested in developing technology-based solutions to major societal problems, and seriously considered transforming NASA into a general applied science agency—a 'new NASA'—to take on that responsibility."[45]

The idea had come to Nixon when someone told him that new technologies could desalinate seawater and use it for agriculture or as drinking water. The president became obsessed with the idea and seriously considered renaming NASA, reasoning that if the agency could land humans on the moon, it could also tackle terrestrial problems.

The following is an excerpt from a meeting between Nixon, his chief advisor John Ehrlichman, and Secretary of the Treasury George Shultz:

> **Nixon:** Can we name it something other than National Aeronautics and Space?
> **Ehrlichman:** We're working on that.
> **Nixon:** If we put some research projects in a few places, wonderful. Put a lot of them in California [Nixon had won the 1968 California election by a margin of three percentage

> points over his Democratic opponent, Hubert Humphrey, following a Democratic victory four years earlier. For him, jobs and prestigious projects in the important state of California were crucial to repeating this success in the 1972 election, R.Z.].
>
> **Shultz:** Why not take full advantage of everything about this? In broadening NASA's horizons, we can finally do that. They like the idea of a well-defined mission in space and aeronautics, but they are gradually being brought to think a little bit more broadly.
>
> **Nixon:** We can put it in terms of taking them to a mountaintop. We bring them in, we say, look, you have shown how it can be done, in other words we give you a project and we say go off and do it. Now we're going to give you this one [desalination], and you go out and do it. And that's the best way to get the teams [working], and you know they get, they go 'Ra-Ra-Ra' and they wear the blue shirts with . . . letters and things.[46]

Ehrlichman wrote to Shultz after the meeting, "the President would like serious consideration given to changing the name of NASA to something designating a more domestic orientation."[47] NASA was assigned the task of investigating what other technological problems, entirely unrelated to space travel, it could solve.[48] Apparently, the successful moon landing made politicians delusional, and they thought that other challenges—especially those with broad voter appeal—could be overcome in a similar way, that is, through government industrial policy and planning.

Consequently, on December 23, 1971, President Nixon announced the "War on Cancer" by signing the National Cancer

Act. He described the initiative as potentially the most significant measure of his presidency. Today, however, it is universally acknowledged that this project failed and represents yet another example of "mission-oriented innovation policies" (MOIPs), which, despite being championed by economists like Mariana Mazzucato, often fall short in practice.[49]

Nixon even considered canceling the long-planned Apollo 16 and Apollo 17 missions and reallocating the funds to other initiatives. This would have solved two problems for Nixon: He was worried that a lunar mission failure would harm his reelection prospects in 1972, and he believed that using the money strategically to have NASA address "terrestrial" issues that resonated more with voters could enhance his election appeal. Rational arguments from NASA scientists, who pointed out that the two missions represented less than half a percent of the Apollo program's total costs but accounted for over 50 percent of productive time on the lunar surface, fell on deaf ears.[50] Other arguments carried more weight, particularly the concern that cutting NASA's budget and canceling planned projects could lead to increased unemployment in California's aerospace industry, thereby jeopardizing Nixon's reelection.

In terms of space exploration, various ideas had been discussed, most notably a project called the "space shuttle." The question of job creation in California was of paramount importance, especially as California held forty-five electoral college votes in the U.S. presidential election, more than any other state. President Nixon directed Caspar Weinberger, who was tasked with the preparation and oversight of the federal budget, to "personally stay on the project of jobs for California and that you make sure that there is no let up in the efforts on this."[51]

The debate over whether the space shuttle should be the focus of space activities following the end of the Apollo program, and on what scale the shuttle should be built, dragged on for almost three years, with the impact of any decision on jobs in California and the upcoming elections in the state playing a central role.[52]

Short-term election tactics had far-reaching and long-term consequences, as Logsdon illustrates: "President Richard Nixon and his associates, with the decision to develop the space shuttle, had finally given an answer to the question: 'What do you do next, after the Moon?' That answer defined much of the U.S. civilian space program for the next 40 years. John F. Kennedy's 1961 decision to go to the Moon led to the Apollo program, which lasted from 1961 to 1975; Richard Nixon's decision to build the U.S. post-Apollo space program around the space shuttle had far more lasting impact."[53]

It is likely that politicians were not fully aware of the long-term ramifications. They were accustomed to thinking in shorter time frames, particularly in the period leading up to an election. That was certainly the case with this particular decision.

The Domestic Council, a White House body that dealt with domestic affairs—essentially the domestic counterpart to the National Security Council—warned that, "Cutbacks in Defense and NASA by 1972 will shrink expenditures by 30 percent from 1968 levels, creating unemployment (850,000 workers)—especially among scientists and engineers (an additional 130,000)." An official from the Domestic Council further noted that, "the unemployment is very localized," with 43.5 percent in the Pacific region, particularly in Los Angeles. The historian Logsdon concludes: "The connection between aerospace employment and the space shuttle, already evident in 1970, was to prove an important factor in the final decision to approve the NASA-preferred shuttle at the end of 1971."[54]

Nixon's advisor Ehrlichman recalls that the issue of the jobs created by the space shuttle project, "was a very important consideration in Nixon's mind. . . . I can recall conversations about that, which were highly persuasive. . . . You must not underemphasize that element, that employment element, in Nixon's decision [on the shuttle, J.M.L.]. . . . When you look at the employment numbers, and you key them to the battleground states [those states with electoral votes important to winning the presidency in the 1972 election, J.M.L.], the space program has an importance out of proportion to its budget."[55]

Nixon himself noted that, "In Florida and California this [approving the shuttle, J.M.L.] is a big deal. It will save the aerospace industry." And Logsdon comments: "Whatever else his approval of the space shuttle meant to Richard Nixon, he saw it as an asset in terms of his reelection prospects."[56]

As previously mentioned, alternate projects had been discussed, including a large-scale endeavor to take astronauts to Mars. However, a mission to Mars was quickly dismissed due to its perceived exorbitant costs. At the time, the space shuttle seemed to be a financially far more viable option, which, as we shall see, was based on a number of false assumptions.

Matthew H. Hersch, Professor of the History of Science at Harvard University, has thoroughly analyzed the history of the space shuttle and concluded: "That NASA would follow project Apollo with some kind of space program was a given among the nation's political leadership in 1972, but arguing most significantly for the shuttle's development was the fact that it was far cheaper than most alternative proposals, especially an Apollo-style mission to Mars. Of the panoply of space vehicles and projects NASA considered undertaking after Project Apollo—including space stations, nuclear orbital ferries, Venus fly-bys, and Mars landings—the shuttle offered the lowest cost and the most immediate utility, although its value hinged upon unrealistic economic assumptions."[57]

The space shuttle's specific objectives—and this is remarkable—were initially unclear. There was a wide variety of ideas and requirements, some of which were mutually exclusive. The eventual configuration of the space shuttle was largely a result of political compromise between various interest groups. The discussions had taken a total of three years.[58] As mentioned, the primary concern for political decision-makers was not the mission's purpose, but rather a desire to convey the message that the United States would remain active in space exploration, at manageable costs, without excessive risk, while ideally creating or at least preserving jobs in crucial electoral states. "Nixon," writes Hersch in his study of the space shuttle, "insisted only that NASA produce *something*, and that, with the 1972 election quickly approaching, it be manufactured in the electoral swing state California."[59]

President Nixon clearly liked the design of the shuttle; in fact, he even decided to keep a model loaned to him by NASA for a press conference without returning it.[60] Besides the anticipated job creation in California, the decision was also influenced by the close ties between Nixon and Willard F. Rockwell Sr., a major donor to Nixon's election campaign, the founder of North American Rockwell, the company responsible for producing the space shuttle, and a personal friend.[61]

Decisions on which parts of the space shuttle should be produced by which companies were not driven by objective considerations, but, as is common with defense projects in the United States, by political criteria. "Companies in congressional districts across the country manufactured the wings, rudder, and other portions of the orbiters to ensure the shuttle's support by a diverse array of lawmakers, and then shipped the parts to Rockwell's plant in Palmdale, California, for final assembly."[62]

NASA initially had the idea of building a reusable rocket to help reduce costs. This was by no means a new idea; there had already been several concepts and attempts, but none had been successful. The idea of full reusability was also soon abandoned, as it was deemed too expensive.[63] Adding to the problems, every stakeholder group had a different vision and different requirements for the space shuttle: the scientists, the military, and those who cared about manned spaceflight. Hersch draws a bleak conclusion: "The shuttle's multiple identities—a human spacecraft, a carrier for satellites, an orbital laboratory, a craft to service space stations and interplanetary ships, a delivery vehicle for commercial payloads, a tool of diplomacy and international cooperation, and eventually, a military spaceplane—were always in conflict, and as each required a different machine, NASA could never satisfactorily reconcile them. Within a few short years, the shuttle had transformed from an optimistic effort to create a spacefaring civilization to an awkward, winged vehicle that satisfied no one, including Air Force planners whose interest in the shuttle increasingly determined its design."[64]

The space shuttle, it could be said, "was an artifact of a confused and chaotic political age. Like other big technologies of the period, the shuttle epitomized the crisis- and consensus-driven nature of twentieth-century American politics and the tendency to build large systems by committee."[65]

Many involved, both at NASA and in the Air Force, harbored doubts regarding the concept of a "jack-of-all-trades" spacecraft that launched like a rocket and landed like an airplane. The notion of constructing a spacecraft with wings, which would primarily serve as a hindrance for most of the journey and only be necessary for landing, left many engineers skeptical from the outset. Ultimately, the space shuttle was a political compromise designed to address numerous, at times conflicting, interests and requirements. Unlike the Apollo project, which had the clearly defined goal of landing humans on the moon, the space shuttle was envisioned as an "all-purpose space vehicle"[66]—without a clearly defined objective.

In many instances, entirely unrelated motives influenced decisions. For example, Rockwell was chosen to manufacture the spacecraft because President Nixon wanted to award the contract to a company based in the then swing state of California before the election.[67] "Compromise and thrift had driven its configuration, and its principal benefit lay in its authorization in an election year."[68]

Contracts were frequently granted based on questionable political considerations. Another example was the decision to award the contract for the space shuttle's solid rocket boosters (SRBs) to Thiokol, a decision that "remained clouded by accusations of favoritism by Utah politicians and civil servants."[69] The *Challenger* disaster of January 1986 would never have happened if Thiokol's management had not made the erroneous decision to authorize the launch. This was all the more tragic because the launch went ahead despite an employee's urgent warning that *Challenger*'s O-rings could lose their elasticity in the unusually cold weather, which is exactly what happened and led to the accident in which all seven astronauts died.

In order to enhance the shuttle's public relations, one of the crew was a teacher, who died along with the six astronauts—effectively halting the prospect of private space tourism on the shuttle, at least for the time being. Objectively, the shuttle was no riskier than the Apollo spacecraft. The fact that more people died on the shuttle was simply a result of there being seven astronauts on board instead of three. Despite some claiming that flying in the space shuttle was as safe as flying in an airplane, professional astronauts knew perfectly well that this was not true.

The initial response to the *Challenger* disaster was an immediate ban on the shuttle carrying commercial payloads. Space expert Rand Simberg comments on this decision: "This was actually a good policy, because the shuttle had from its inception retarded the development of commercial launch vehicles by allowing a taxpayer-subsidized system to undercut the cost of any potential competitor. This (for obvious reasons) inhibited investment in same, and the decision to end that policy did enable the development of the commercial launch industry as it exists today. But as is often the case, particularly with government, it was an example of a good policy occurring for a really dumb reason. The policy wasn't changed to create a commercial launch industry—it was changed because people said that astronauts *shouldn't be risking their lives for the mundane task of launching satellites.*"[70]

And yet, similar scenarios unfold millions of times around the world every single day: Truck drivers risk their lives transporting goods and, in the United States alone, several hundred die doing so each year.

From an economic standpoint, the space shuttle program was a complete disaster. The development of the rocket was originally forecast to cost between $4.5 and $5 billion. According to different estimates, the running costs per flight were between $5 million[71] and $10 million.[72] In reality, politicians were hardly interested in the running costs, since these would continue even after Nixon left office.[73]

The projected costs were reduced by continuously increasing the projected number of flights per year, from an initial eight to twelve flights per year to up to 140 flights per year, i.e., about three times a week.[74] The logic is simple: the higher the number of calculated flights, the cheaper each flight would be. In reality, however, the shuttle only flew 135 times during its thirty years of service from 1981 to 2011.[75] This was partly due to the long hiatuses following the two accidents, but also partly because the costs and effort required for repairs after each flight were far greater than expected. It was by no means the case that—as had been predicted—the shuttle could simply land, have a quick inspection, and take off again, like an airplane. After a space shuttle flight, maintenance and repairs typically took three to six months and involved several hundred to over 1,000 technicians. Inspecting and repairing the heat shield, as well as completely disassembling and testing the main engines, proved particularly complex. The heat shield consisted of over 24,000 individual ceramic plates, each uniquely shaped and numbered. After every flight, these plates had to be carefully checked and aligned, and any damaged parts had to be laboriously replaced, making maintenance both time-consuming and expensive.

According to NASA reports and analyses, the cost of a space shuttle launch, including operation and infrastructure, was approximately $1.5 billion, delivering up to 27,500 kg of payload into low earth orbit (LEO)—roughly $54,500 per kg.[76] The total cost of the shuttle program exceeded $200 billion by 2011[77] (equivalent to approximately $300 billion in 2025 prices).

On February 1, 2003, the second fatal space shuttle accident occurred when *Columbia* broke apart during reentry over Texas. All seven crew members perished. That same day, an investigative committee, the *Columbia* Accident Investigation Board, was established. The board's findings were devastating, revealing that the shuttle had fallen short of all initial expectations, yet it remained essential: "The shuttle has few of the mission capabilities that NASA originally promised. It cannot be launched on demand, does not recoup its costs, no longer carries national security payloads, and is not cost-effective enough, nor allowed by law to carry commercial satellites. Despite efforts to improve its safety, the shuttle remains a complex and risky system that remains central to U.S. ambitions in space."[78]

One of the most significant negative consequences was that the government-run space shuttle program long hindered the development of a private rocket industry, according to Hersch: "Instead of exploring space, NASA would replace the embryonic free market for launch services with a single government provider that purchased expensive, unreliable rockets from key defense contractors selected by political appointees, and then priced their flights below cost for favored users, destroying the competitive pressure that might have improved the technology."[79]

Despite its efforts, NASA had failed to develop an alternative launch vehicle to replace the shuttle. Therefore, although the end of the shuttle program was announced in January 2004, eleven months after the accident, twenty-two more flights took place after a two-and-a-half-year hiatus, beginning in July 2005. Finally, the last shuttle flight on July 21, 2011—the 135th flight in the entire history of the shuttle program—brought the program to a close.

We already know what happened next. Since the United States couldn't send its astronauts to the ISS using a trampoline, it was dependent on Russian Soyuz rockets for years to come.

Perhaps it would have been better to listen to the American economist William Niskanen, who, in October 1971, proposed a radical alternative during a meeting of the Office of Management and Budget, namely "to have the private sector, using its own financial resources, develop the next generation space transportation system, and then sell transportation services to NASA and the Department of Defense to recoup that investment."[80]

However, George M. Low, the deputy administrator of NASA, reporting directly to then-Administrator James C. Fletcher, countered: "The reason for not doing it is that it simply won't work: if the idea is to cancel the space program, this might be a way to do it."[81] As things turned out, he was greatly mistaken. But he was not alone. His statement is just one of many examples of politicians and government officials grossly overestimating their own role and capabilities while underestimating the power of the market.

The history of manned spaceflight after Apollo reads like a textbook proof of public choice theory, which applies economic principles to political behavior to explain how political decisions are made. The pursuit of space exploration, or even scientific advancements, took a back seat to considerations such as enhancing politicians' public image, appeasing lobbyists, and, of course, securing the most votes in key states. Only someone who naively assumes that politicians are primarily interested in the common good and maximizing prosperity could be surprised by this finding.

CHAPTER 2

THE SECOND SPACE AGE

In March 2020, the United States implemented its first measures to counter the COVID-19 pandemic. On April 1, Florida Governor Ron DeSantis declared a statewide lockdown—schools, bars, and nightclubs were closed due to concerns over the coronavirus. But Douglas Hurley had other concerns. Born in 1966 in Endicott, New York, the birthplace of the tech giant IBM, Hurley was preparing for his third space mission. Space exploration was his passion, and he would spend more than ninety days in space before his retirement from NASA in 2021. His wife was also an astronaut.

Hurley had first flown to the International Space Station (ISS) on the space shuttle in 2009 and was also aboard the historic final flight, STS-135, which launched on July 8, 2011. Now, on May 27, 2020, he had taken his seat on the Crew Dragon spacecraft, and was set to fly to the ISS on a Falcon 9 rocket from the private company SpaceX. Elon Musk had named the Dragon spacecraft after the 1960s song "Puff, the Magic Dragon" in response to critics, who repeatedly declared that his business plan would never succeed.[82]

This mission marked the first time in decades that a new American rocket would carry astronauts into orbit, with the last such debut being the space shuttle four decades earlier. Most importantly, it would be the first crewed launch into Earth orbit by a private space company and the first crewed launch from U.S. soil to transport astronauts to the ISS since the end of the space shuttle program in 2011. Although the mission was conducted on behalf of NASA, the launch system was entirely developed by SpaceX, comprising the Falcon 9 rocket and Crew Dragon spacecraft.

Back in March 2019, the Crew Dragon spacecraft had embarked on its inaugural uncrewed test flight to the ISS. On board, however, its sole "passenger" was a test dummy affectionately named Ripley, in homage to the character from the film *Alien*. The mission was designed to test critical safety systems, including life support, under realistic conditions. Ripley was outfitted with sensors to gather data on acceleration, pressure, and vibrations throughout the flight. The successful mission was a crucial milestone in progressing toward the first crewed mission with Crew Dragon.[83]

As he boarded the spacecraft, Hurley wasn't sure it would actually lift off. His first space shuttle flight in 2009 had faced multiple delays due to technical problems, and it wasn't until the sixth attempt that it finally succeeded. "Because of that, whenever I go to the pad, I assume we're going to scrub," he said.[84]

This time was no different. The weather didn't play ball and Hurley and his copilot Robert Behnken were forced to abort the mission, leaving everyone disappointed. Elon Musk wanted to speak to the Launch Weather Officer who had made the decision to abort. Although an earlier thunderstorm had passed, the launch was halted sixteen minutes before liftoff due to the "field mill rule," which had been in place ever since an incident during the Apollo

12 launch. This rule states that if an electrostatic field strength sensor (field mill) is activated too close to the launch site, there is a risk of lightning being generated during the launch, posing a threat to both the rocket and its crew. NASA and SpaceX are required to stop any launch countdown as soon as the field strength exceeds the set threshold: an automatic "no-go" to avoid the risk of launch-induced lightning.

Musk, known for his impatience with what he sees as the many nonsensical government regulations from NASA and environmental agencies, accepted the regulation's rationale after weather officer Mike McAleenan provided a detailed explanation. "If you give Elon the straight answers, with all of the detail he wants, he's very happy," the weather officer said. "But don't try and bullshit him, because he'll smell that out."[85]

U.S. President Donald Trump had come to watch the launch and bask in Musk's success—a positive diversion from all the negative coronavirus news. However, when the launch was aborted, Trump didn't understand the reason—and didn't want to understand. Instead, he blasted NASA Administrator Jim Bridenstine in front of the crowd of VIPs assembled for the launch.[86]

Three days later, on May 30, the launch was successful, and nine minutes and 31 seconds later, the first stage of the Falcon 9 rocket landed safely on the recovery vessel "Of course I still love you." Ten million people watched the launch live on streaming platforms. Musk then drove with his girlfriend Grimes, his brother Kimbal, and a few others to a vacation resort in the Everglades, a two-hour drive south of Cape Canaveral. It took a while for everyone to fully grasp the historic moment, and, at one point, Kimbal jumped up and shouted: "My brother has just sent astronauts up into space!"[87]

Hurley and Behnken spent nearly sixty-four days aboard the ISS before they splashed back down to earth in their Dragon capsule off the coast of Pensacola, Florida, on August 2. Hurley's first request was pizza, which he duly received. He later recalled: "It was the second space age. And it started in 2020."[88]

But how did it all begin?

Elon Musk, born on June 28, 1971, in Pretoria, South Africa, had been interested in space travel since he was a boy, but it was in the early 2000s that his interest intensified. He recalled visiting the NASA website to study their plans for a Mars mission. "I figured it had to be soon, because we went to the moon in 1969, so we must be about to go to Mars," thought Musk.[89] However, to his dismay, he found there were no specific plans at all. Musk's online research led him to an announcement for a dinner in Silicon Valley, hosted in August 2001 by the Mars Society, an organization dedicated to the manned exploration and colonization of the Red Planet. Tickets cost $500, but Musk sent a check for $5,000, a deliberate gesture to capture the attention of Mars Society founder and president Robert Zubrin.

Zubrin was indeed impressed and invited Musk for a personal meeting. Shortly thereafter, Musk spent a day with Zubrin at his company near Denver. This meeting marked the beginning of Musk's deeper involvement in space exploration. Later that same year, he donated $100,000 to support the Mars Desert Research Station. For a time, Musk even served on the board of the Mars Society.[90]

Musk was looking for someone who could help him turn his vision into reality. Zubrin suggested he contact rocket engineer Jim Cantrell. Cantrell later recalled his surprise at receiving a call from a stranger who introduced himself as an internet millionaire named

Musk, who said he had gotten the number from Robert Zubrin, the founder of the Mars Society: "I am the founder of Zip2 and PayPal and believe that mankind has to become a multi-planetary species to survive, and I want to do something with my money to show that this is possible."[91]

Cantrell didn't know what to think. Was it a crank call? When the call came in, he was driving his convertible with the top down, so he asked if he could call back in fifteen minutes. As soon as he was home, Cantrell checked to see whether a company called PayPal actually existed. Once he was satisfied that the call was legitimate, he tried to call back, but instead of Musk's voice, Cantrell heard the beeps of a fax machine. Shortly afterward, Musk called him again and they arranged to meet at the airport along with Zubrin.

Musk outlined his plan to send a capsule containing a crew of mice into Earth orbit, where the capsule would rotate to simulate an environment of Mars gravity and allow them to collect data on how mammals from Earth would respond to Mars gravity, and how children born in Mars gravity would develop.[92] Due to the high cost of U.S. rockets, Musk was weighing up whether to buy Russian rockets, and asked Cantrell to make the contacts for him. Cantrell was not particularly keen on the whole mouse experiment idea. The next plan they came up with was to send a rocket to Mars and show that a plant could grow in a small greenhouse on the Red Planet.

At the beginning of November 2001, Musk, Cantrell and the later NASA director Mike Griffin flew to Moscow. "It's important to realize that Elon was, at the time, a twenty-something 'kid' from Silicon Valley who had done very well financially and dressed much like the other entrepreneurs in Silicon Valley. He dressed poorly in the eyes of the Russians, which was something of great import to them. I had warned Elon about this and how the Russians would judge him based on his appearance."[93]

When Musk shared his ambitious vision of establishing human colonies on Mars, one of the Russians was so annoyed that he spat on Musk and Cantrell's shoes. They met with other Russians to negotiate a deal for a rocket, but these talks were also unsuccessful.

On the flight back from the failed negotiations, Cantrell saw Musk working on his laptop: "Hey guys, I think we can build the rocket ourselves,"[94] Musk declared. The others laughed at him at first. But a quick glance at Musk's calculations convinced the rocket engineer that Musk's idea made a lot of sense and he had clearly immersed himself in the subject matter, extensively studying books on fuels and rockets.

At a meeting with some former employees of PayPal, the company Musk had founded and later sold, becoming rich in the process, Musk was sitting by the pool, engrossed in a tattered manual for a Russian rocket engine. When one of his friends asked what his plans were for the future, Musk answered: "I'm going to colonize Mars. My mission in life is to make mankind a multiplanetary civilization." His former colleague's reaction? "Dude, you're bananas."[95]

I have extensively studied successful individuals who set ambitious goals, yet no goal is perhaps as grand or seemingly "impossible" as Elon Musk's aspiration to reach Mars. What drives Musk? Why is he so determined to colonize Mars? Before we examine the next steps for Musk and SpaceX, it is essential to understand the motivation behind his goal. Without a compelling "why," Musk would not have persevered through so many setbacks and defeats. And it is a fact that every decision he makes, down to the smallest detail, is based on whether it brings him closer to his ultimate goal:

"The lens of getting to Mars has motivated *every* SpaceX decision," Musk stated.[96]

Especially when things got difficult, Musk would always remind his team of the overarching goal of conquering Mars. Every week, he held a meeting called "Mars Colonizer," where he shared his vision for a Mars colony and discussed how it should be governed. "We tried to avoid ever skipping Mars Colonizer, because that was the most fun meeting for him and always put him in a good mood," recalls one of Musk's former assistants.[97] "They're sitting around seriously discussing plans to build a city on Mars and what people will wear there," one employee wondered, "and everyone's just acting like this is a totally normal conversation."[98]

His biographer, Walter Isaacson, notes that Musk has cited three primary reasons for his ambition to reach Mars:

First, progress is not inevitable and can come to a standstill. History has repeatedly shown periods of stagnation or regression. "Do we want to tell our children that going to the Moon is the best we did, and then we gave up?"[99] he questioned, pointing to the stagnation in manned spaceflight since the early 1970s.

Second, Musk wants to settle millions of people on Mars, because he believes it is humanity's only long-term chance of survival. Earth has been repeatedly struck by asteroids—and as we know, this once led to the extinction not only of the dinosaurs, but of almost all life on Earth.

Some authors criticize the concept of colonizing or terraforming Mars as "pure science fiction": "The argument that emigration is necessary to avoid being wiped out by an asteroid is nonsense. Such objects also hit Mars."[100] Of course, anything is possible, but the likelihood of both Earth and Mars being hit by a large asteroid at roughly the same time is practically zero. Musk is right when he says:

"If we are able to go to other planets, the probable lifespan of human consciousness is going to be far greater than if we are stuck on one planet that could get hit by an asteroid or destroy its civilization."[101]

And third, Musk believes that people need big, inspiring goals. In fact, he sees this as the meaning of life. At a conference in June 2016, he said the following about his motivation: "There are a lot of negative things in the world. There's a lot of terrible things happening all the time. There are lots of problems that need to get solved—lots of things that are miserable and kind of get you down. But life cannot just be about solving one miserable problem after another. That can't be the only thing. There need to be things that inspire you—that make you glad to wake up in the morning and be part of humanity. It is time to go forth, become a star-faring civilization, be out there among the stars, expand the scope and scale of human consciousness. I find that incredibly exciting. That makes me glad to be alive."[102]

It is probably impossible for people who don't set themselves ambitious goals and are content with the status quo to understand Musk's motives. The great German spaceflight pioneer Hermann Oberth wrote in his 1957 book *Man into Space*: "And what would be the purpose of all this? For those who have never known the relentless urge to explore and discover, there is no answer. For those who have felt this urge, the answer is self-evident."[103] In a similar vein, in 1923, when the British mountaineer George Herbert Leigh Mallory was asked during a lecture tour of the United States why he wanted to climb Mount Everest, he simply replied: "Because it's there."[104]

To understand Musk, one must always bear in mind that he is driven by this grand goal, but also that he feels an immense sense of urgency. His thinking is bigger than that of managers in traditional aerospace companies, and his timelines are far more ambitious. As Ashlee Vance writes in his biography of Musk: "In regards to time, Musk may well set more aggressive delivery targets for very difficult-to-make products than any executive in history."[105] One employee reported: "He will pick the most aggressive time schedule imaginable assuming everything goes right, and then accelerate it by assuming that everyone can work harder."[106] When Musk was estimating how long a software project would take, the employee recounted, half-jokingly, he would start by timing the seconds needed to write a line of code and then extrapolate that out to however many lines of code he expected the final piece of software to have. "Everything he does is fast. He pees fast. It's like a fire hose—three seconds and out."[107]

According to Musk's biographer Walter Isaacson, Musk describes his own work ethic as follows: "A maniacal sense of urgency is our operating principle."[108] It's a phrase he has repeated again and again and again. For example, when one of his employees was working on the new Merlin engines for the Falcon 9 rocket and presented an aggressive schedule for completing one of the versions, it wasn't aggressive enough for Musk, reports Isaacson. "How the fuck can it take so long?" Musk asked. "This is stupid. Cut it in half!"[109] Musk is aware that reaching Mars is an almost superhuman goal, but he also believes it depends on *him*; he sees the colonization of Mars as *his* great mission.

But, as the age-old saying goes, even the longest journey begins with a single step. It would be impossible to trace the evolution of SpaceX in detail here, but it is important to understand its interplay with NASA. For decades, NASA had made no progress in manned spaceflight. With the end of the space shuttle program, even NASA realized something had to change and that it could not move forward without the help of private companies.

More and more experts had also acknowledged the need for a paradigm shift, including the participants of a conference held by the Cato Institute in March 2001, the results of which were published in 2002 in the book *Space: The Free Market Frontier*—coincidentally, the same year Elon Musk founded SpaceX. This book foreshadowed, conceptually, what would soon become NASA policy.

The question posed at the conference was: "What has happened in the past three decades to delay humankind's full exploitation of space, and what can be done to change the situation?"[110] The main obstacle, according to Robert W. Poole Jr., founder of the Reason Foundation, was, "the central planning approach: the assumption that engineers and government planners can devise the one best way to launch payloads to space . . . and that it is simply a question of pouring enough funding into the chosen model for long enough to make it succeed."[111]

Buzz Aldrin, the lunar module pilot for the first human landing on the moon, criticized: "The fundamental building block of the U.S. space program is the transportation capability that provides access to space. With the exception of the space shuttle, American space access capabilities have changed little in the past four decades and no progress has been made in solving the greatest obstacle to space development—the high costs of space access."[112]

Alternative projects NASA was working on had also proved unsuccessful, including the X-33 and X-34, which together cost more than $1 billion. The X-33 was an experimental spaceplane developed by NASA and Lockheed Martin in the 1990s as a prototype for a reusable space transportation system called VentureStar. The project was canceled in 2001 before its first flight. The X-34 was an unmanned hypersonic vehicle developed by NASA in the 1990s to test cost-effective reusable spaceflight technologies. However, after successful ground tests and several delays, the project was also terminated in 2001. "X-33 and X-34 both demonstrated that NASA has a less-than-stellar track record in picking the right technologies,"[113] complained Marc Schlather, director of the Senate Space Transportation Roundtable.

What alternatives did the conference come up with? Robert W. Poole suggested that, "instead of defining in great detail the specifications of a new launch *vehicle* . . . these government agencies would simply announce their willingness to pay $X per pound for payloads delivered to, say low earth orbit (LEO). In other words, instead of the typical government contracting model, which has failed to change the cost-plus corporate culture of aerospace/defense contractors, NASA and the other government agencies with space transportation needs would purchase *launch services*."[114]

Doris Hamill, Philip Mongan, and Michael Kearney from the company SpaceHab called for a paradigm shift in space commercialization in their 2002 article "Space Commerce: An Entrepreneur's Angle" and correctly predicted: "This approach to attracting commercial users does not require the space agencies to perform market development activities, to command its contractors to find efficiencies that will undercut the contractor's revenue stream or to establish limits on how much they will subsidize commercial research. They only need to agree to purchase commercial services that meet their research needs within their budgets. The rest will happen by itself."[115]

And that's exactly how it played out: In January 2006, NASA announced a new strategy known as COTS (Commercial Orbital Transportation Services), a funding program aimed at enlisting the assistance of private companies in transporting equipment, supplies, and experimental devices to the ISS. Instead of conducting its own missions to the ISS, the concept behind COTS was to delegate the task of carrying out missions to the ISS to private companies with their own spacecraft. By 2010, NASA planned to allocate a total of $500 million to this initiative, much less than the cost of a single space shuttle flight. Unlike previous NASA projects, the spacecraft were to be entirely financed, constructed, and operated by the private companies.

Finally, this shift in strategy signaled NASA's acknowledgment that its previous way of collaborating with private companies, which had proven successful during the Apollo program, had failed in the decades that followed. The previous approach can only be described as a centrally planned form of "cooperation," with private companies having effectively functioned as underlings of a government agency.

To appreciate the significance of this change, it is essential to consider NASA's previous mode of operation: The agency had historically contracted private industry to construct rockets or probes. However, as noted by Joan Lisa Bromberg in her seminal work "NASA and the Space Industry," NASA's original philosophy was to maintain complete control: "It was government acting in the public interest that had to determine what should be done, when it should be done, and for how much money" And: "Federal agencies intrude into the activities of their private-sector contractors far more than the uninitiated might intuit, given the free enterprise structure of the U.S. system."[116] NASA even intervened in the companies' personnel policies, their organization, the selection of their contractors and subcontractors, etc.[117]

All of this worked so long as virtually unlimited resources were available. At the height of the Apollo program, more than 400,000 people and 20,000 companies and universities were involved in the project—"evidence of a warlike mission,"[118] as Christopher Wanjek writes. Whether money was wasted or whether any kind of efficient work was being done was irrelevant; after all, it was a matter of paramount national importance. The United States was determined, at any cost, to win the race to the Moon.

The so-called cost-plus system was particularly absurd. NASA's contractors were required to meticulously document their expenses, and were then permitted to add a moderate profit margin of approximately 8 to 10 percent to the overall cost. In a market economy, companies always strive to keep costs as low as possible and increase their profits through efficiency gains. According to the absurd logic of cost-plus contracts, however, a company's profit was higher the more costs it incurred.

Space expert Robert Zubrin writes: "No farmer or maker of widgets would ever staff up his or her operation with four administrative personnel for every field or factory worker. At major aerospace companies, this is done all the time. As a result, it is the norm for such contractors to have overhead rates exceeding 300 percent. Indeed, at the Martin Marietta company (later Lockheed Martin), where I was employed from the late 1980s through mid-1990s (and which was, along with Boeing, one of the two most successful of the eight major aerospace companies of the era), we at one point had more than 13,000 people at our primary facility, with fewer than 1,000 working in the factory—leading one wit to scoff: 'At Martin Marietta, overhead is our most important product.' Clearly, such a system is inimical to cost reduction. The government could probably cut its procurement costs in half nearly immediately, and perhaps as much as tenfold eventually, simply by getting rid of this insane system and buying things the way everyone else does."[119]

The government had recognized that it was incapable of building a rocket itself and that private companies would have to take on the task. However, the companies would not be allowed to make profits deemed "excessive" by the authorities. That was, at least, what politicians and civil servants had decided. In reality, however, the outcome was quite different. The companies ended up producing rockets at exorbitant cost and with excessively long production times. It almost became a ritual: The companies would declare that, unfortunately, everything was much more complicated, would take much longer, and would cost much more than originally expected. This was the easiest way to boost profits. Instead of preventing the companies from generating excessive profits, the government found itself having to pay even more.

But it didn't stop there. NASA dictated to the companies, down to the minutest detail, how their rockets should be constructed. Consequently, the companies functioned as nothing more than implementing bodies for a government agency. In doing so, the government deprived itself of the creativity of the many specialists within these private companies.

The COTS program was a desperate, last-ditch attempt to chart a new course after the failure of the space shuttle and all previous attempts to develop a replacement. NASA now shifted its focus from the procurement and production of rockets to purchasing services, such as transporting cargo or people to and from the ISS.[120]

This perfectly reflected Elon Musk's entrepreneurial mindset, as his then-employee John Couluris explained in a conversation with NASA, as reported by Eric Berger: "He [Couluris] tried to explain SpaceX's then unorthodox philosophy. Musk did not want to build a spacecraft and sell it outright to NASA. Rather, he wanted to build a spacecraft and charge NASA a fee to fly its cargo. It's like FedEx, Couluris explained. You provide us a package, and we'll deliver it to space for you."[121]

It took a while for NASA to understand this philosophy and be ready to fully implement it. As Couluris explains, "This seems obvious today, but the look of horror on their faces was very, very real. They were like these guys, and that guy from PayPal, are going to approach our $100 billion station with six astronauts onboard?"[122]

Hardly anyone in Washington, D.C., was comfortable with the idea of private companies transporting NASA astronauts to the ISS. Space expert Eric Berger reports: "Only government space agencies were thought capable of such a sacred task. Even the architect of the Commercial Cargo program, NASA-Administrator Mike Griffin, viewed crew flights as a bridge too far."[123]

Most were even more skeptical. In 2010, Republican U.S. Senator Richard Shelby, who wielded considerable influence over NASA's budget as budget secretary, criticized efforts to solve NASA's problems by relying more heavily on private companies like SpaceX, calling it a "death march" for NASA: "We cannot continue to coddle the dreams of rocket hobbyists and so-called 'commercial' providers who claim the future of U.S. human spaceflight can be achieved faster and cheaper."[124]

These were strong words, especially as NASA's shuttle program had fallen far short of its stated objectives and that each shuttle launch cost approximately $1.5 billion. They were also strong words considering that launch costs had remained largely stagnant from 1970 to 2010, a period during which all other sectors experienced substantial productivity gains. Furthermore, NASA had wasted over $1 billion on several unsuccessful attempts to develop reusable rockets, such as the X-33 and X-34. And these were strong words bearing in mind that, since the end of the space shuttle program, the United States was utterly dependent on increasingly costly and outdated Russian rockets to reach the International Space Station.

However, the potential for alternative approaches had long existed in the political arena. Notably, President Ronald Reagan, in a visionary speech in 1984, foresaw a promising future for private spaceflight: "The third goal of our space strategy will be to encourage American industry to move quickly and decisively into space. Obstacles to private sector space activities will be removed, and we'll take appropriate steps to spur private enterprise in space. We expect space-related investments to grow quickly in future years, creating many new jobs and greater prosperity for all Americans. Companies interested in putting payloads into space, for example, should have ready access to private sector launch services. . . . So, we're going to bring into play America's greatest asset—the vitality of our free enterprise system."[125]

Several months after this speech, Reagan signed the Commercial Space Launch Act of 1984, allowing private companies to commercialize space exploration and space technology. Nevertheless, it took more than two decades for his vision to become a reality. In 2004, at the behest of President George W. Bush, the Aldridge Commission—officially the President's Commission on Moon, Mars, and Beyond—was established and named after its chairman, Edward "Pete" Aldridge. The commission was tasked with developing strategies for the United States to implement the then-new space program "Vision for Space Exploration," which included long-term missions to the Moon, Mars, and beyond.

Among the recommendations in the commission's final report was a call for greater collaboration with the private sector. Once again, here were experts in their fields asserting that NASA should transition to purchasing services from private companies and that the role of the federal space agency, "must be limited to only those areas where there is irrefutable demonstration that only government can perform the proposed activity."[126] This principle was then enshrined in the 2004 Commercial Space Launch Amendment Act, which declared that the state should, "refrain from conducting activities with commercial applications that preclude, deter or compete with U.S. commercial space transportation activities, unless required by national security."[127]

All of this sounded good, but, despite decades of politicians repeatedly declaring that the private sector should play a larger role and that NASA needed to change, nothing actually happened. Ironically, it was a U.S. president who had little affinity for either capitalism or space travel—the Democrat Barack Obama—who made significant strides in this direction. Obama wasn't interested in spaceflight and ended the costly government-funded Constellation program, which envisioned a government-run return to the moon. He wanted to redirect resources toward expanding the welfare state, especially his pet project, "Obamacare," a government-run healthcare system. In 2015, under Obama, the United States spent one hundred times more on social programs ($1.92 trillion) than on NASA ($18 billion).

But of course, space exploration couldn't be entirely ignored, and when his advisors explained to Obama that private companies should be more involved because they were more cost-effective, he agreed. Sometimes it's better if politicians aren't interested in a topic at all.

The groundwork for the development of the private space industry can be attributed to visionaries such as Peter Marquez, who held several government roles, including Director of Space Policy for the National Security Council. Marquez advised both the George W. Bush and Obama administrations on space policy. "When I was writing the National Space Policy [for the Obama administration]," Marquez said, "I kept a copy of Reagan's first space policy on my desk. I emphasized what commercial industry could do. Since then, we had gone away from all these things that were supposedly hard and fast rules of the American ethos: trust industry, trust capitalism, trust technology. In 2010, I didn't think I was doing anything revolutionary. I was just going back in time to the 1980s."[128]

In a speech at the Kennedy Space Center in 2010, Obama declared: "By buying the services of space transportation—rather than the vehicles themselves—we can continue to ensure rigorous safety standards are met. But we will also accelerate the pace of innovations as companies—from young start-ups to established leaders—compete to design and build and launch new means of carrying people and materials out of our atmosphere."[129]

However, this new form of collaboration and restructuring at NASA could not simply be mandated from the top down. After all, there were numerous skeptics eager to demonstrate that these ideas could not be successfully implemented in practice.

The eventual success of this collaboration also depended on the advocacy of NASA employees like Kathy Lueders, who resisted attempts by middle management to impose additional rules and requirements on private companies. While NASA had formulated over 10,000 requirements for the space shuttle, SpaceX's Dragon ended up with about 400.[130] Some NASA employees recognized that excessive bureaucratic regulations could severely delay, if not stifle, entrepreneurial initiatives. Moreover, Musk has always been particularly sensitive to excessive government regulation.

Three or four times a week, someone at NASA would come to Lueders and tell her, "I'd hate to have your job."[131] Hardly anyone believed that SpaceX would succeed. "But Lueders," Berger writes, "understood, that NASA had no choice. 'This was our Hail Mary pass,' she said."[132]

Ultimately, a productive partnership blossomed between NASA and SpaceX, largely due to the efforts of Gwynne Shotwell, the president and chief operating officer of SpaceX. Shotwell was chosen by Musk because she compensated for his own shortcomings. Musk couldn't have put up with the constant need to negotiate directly with the authorities and NASA officials, whose entire way of thinking went against his grain. Shotwell, however, not only shared Musk's vision but also possessed the patience and diplomatic skills necessary to collaborate effectively with NASA.

The combination of Musk's determination and Shotwell's patience ultimately paid off on May 31, 2020, when astronauts Douglas Hurley and Robert Behnken successfully docked with the ISS. This milestone marked a triumph for private spaceflight, disproving even the most vehement skeptics who had long denied that private companies were capable of achieving such feats.

SpaceX had developed the Crew Dragon capsule in record time, having only begun work on the capsule after NASA awarded the contract. The design of the Crew Dragon, initially known as Dragon V2, was unveiled in May 2014. Just six years later it was ready for the test flight with Behnken and Hurley. Initially, the project involved only Musk and a handful of engineers, but it eventually expanded to include up to one hundred personnel.[133]

Robert Zubrin summarizes the lessons learned during the development of the Dragon capsule: "It is possible for a well-led entrepreneurial team to do things that, previously, it was thought only the governments of superpowers could do, and furthermore, do it in one-third the time and at one-tenth the cost or less than was previously deemed necessary."[134] While SpaceX received $2.6 billion from NASA to develop the Crew Dragon capsule, Boeing received $4.2 billion for its Starliner, which was designed to meet exactly the same NASA requirements.[135]

However, Boeing took an additional four years to send astronauts to the ISS, launching its Starliner for the Crew Flight Test on June 5, 2024. But during its mission to the space station, Boeing's Starliner encountered several issues, including helium leaks and the failure of five out of twenty-eight attitude control thrusters. After thorough analysis, NASA decided it would be too risky to return astronauts Butch Wilmore and Suni Williams aboard the Starliner. Consequently, the Starliner returned to Earth uncrewed on September 6, 2024. Wilmore and Williams remained on the ISS and were returned to Earth on March 18, 2025, by the SpaceX Crew 9 mission (Crew Dragon "Freedom"). Instead of the original eight days, they were forced to remain on the space station for more than nine months. Wilmore, at least outwardly, took it calmly and with humor. At the press conference, he said:

"Stuck? OK, we didn't get to come home the way we planned. But in the big scheme of things, we weren't stuck. We planned and trained."[136]

Fortunately, as things turned out, Boeing's lobbyists, who had argued to NASA that they should be awarded the contract on an exclusive basis as they needed all the funding for themselves, did not prevail. The majority of NASA decision-makers had found Boeing's arguments to be quite persuasive; initially, there had been almost unanimous support for Boeing, even though SpaceX's bid was 40 percent lower than Boeing's. Some of these decision-makers were the same people who had been involved in the space shuttle program.[137] SpaceX almost missed out, which would have meant that even today, U.S. astronauts would still be relying on Soyuz rockets for transportation to the ISS.

The Crew Dragon story serves as a testament to SpaceX's superiority over traditional NASA contractors. But that wasn't all. Five years before the first Crew Dragon flight, SpaceX had achieved something that had eluded NASA—and all other national space agencies—for decades: Three days before Christmas, on December 22, 2015, Falcon 9 made history by becoming the first orbitally launched rocket first stage to successfully land back on Earth—on solid ground at Cape Canaveral. The first launch, reusing a previously flown and landed Falcon 9 first stage, took place on March 30, 2017. This booster stage had been used in April 2016, overhauled, and relaunched, ultimately landing safely once again.

For Musk, however, these milestones were merely stepping stones to his ultimate goal: the development of a super-spaceship capable of taking humans to Mars and back. Before we move onto the story of Starship and its predecessors, christened Falcon 1, Falcon 9, and Falcon Heavy by Musk, in chapter 4, let's first pause for a moment or two. Private space exploration might seem like something completely new but, as we will see in chapter 3, that is by no means the case.

CHAPTER 3

SPACE EXPLORATION WITHOUT GOVERNMENT

In 1949, twenty years before the first moon landing, the world-famous American science fiction author Robert A. Heinlein wrote the short story called "The Man Who Sold the Moon." First published in 1950, the story centers on Delos David Harriman, a brilliant entrepreneur determined to achieve the first manned mission to the moon—regardless of the cost. Harriman is adamant that the mission should not be orchestrated or funded by the government. Instead, he plans the expedition as a private, commercially driven endeavor. With astute business acumen, persuasive propaganda, and adept political maneuvering, he succeeds in attracting investors, overcoming obstacles, and ultimately building a moon rocket.

Initially, his business partners are skeptical, to which Harriman retorts: "You're ask me to show figures on a brand-new type of enterprise, knowing I can't. It's like asking the Wright brothers at Kitty Hawk to estimate how much money Curtiss-Wright Corporation would someday make out of building airplanes."[138]

Harriman is able to demonstrate his track record: He has made a lot of money with his "crazy" ideas so far. But this, he tells the prospective investor, Donald "Dan" Dixon, is the biggest deal in history: "This is the greatest real estate venture since the Pope carved up the New World. Don't ask me what we'll make a profit on; I can't itemize the assets—but I can lump them. The assets are a planet—a whole planet, Dan, that's never been touched! . . . It's like having Manhattan Island offered to you for twenty-four dollars and a case of whiskey!"[139]

Harriman's partner George Strong is worried that a government—the United States or the Soviet Union, perhaps—might claim the moon for themselves. Harriman retorts: "Anyhow, George, if we claim it in the name of the United States, do you know where we will be, as businessmen? . . . We'll be dealt right out of the game. The Department of National Defense will say, 'Thank you, Mr. Harriman. Thank you, Mr. Strong. We are taking over in the interest of national security; you can go home now.'. . . I'm not going to let the brass hats muscle in. I'm going to set up a lunar colony and then nurse it along until it is big enough to stand on its own feet."[140]

Harriman wants to organize the venture privately and explains to his partner: "I am telling you that this is not a government project. If you won't attempt to cope with it on those terms, let me know, so that I can find somebody who will."[141]

By arguing that the Soviets might try to claim the moon, he manages to convince a newspaper publisher who had initially opposed his project. "For years, I've had a recurrent nightmare of waking up an seeing headlines that the Russians have landed on the moon and declared the Lunar Soviet."[142]

He develops countless business ideas to finance his project: "George, you collect stamps, don't you? . . . How much would a cover be worth which had been to the moon and canceled there. . . . I think we could get our Moon ship declared a legal post office sub-station."[143]

Above all, he thinks about real estate deals on the moon. Even before the moon landing, he wanted to start selling land on the moon: "Did you ever read about the Florida land boom, Saul? People bought lots they had never seen and sold them at tripled prices without ever having laid eyes on them. . . . When the boom sags a little we will announce the selected location of Luna City—and it will just happen to work out that the land around the site is still available for sale."[144]

He launches a massive PR campaign, and soon newspapers and magazines are full of headlines like "URANIUM PROSPECTING ON THE MOON—a Fact article about a soon-to-come major industry" or "DIAMONDS ON THE MOON?—A World Famous Scientist Shows Why Diamonds Must Be Common as Pebbles in the Lunar Craters." Space tourism is also set to become a major industry, as reported by one Sunday newspaper: "*Honeymoon on the Moon*—a discussion of the Miracle resort that your children will enjoy, as told to our travel editor."[145]

Heinlein, who started out as a socialist but later became a libertarian, thus developed the idea of a space mission financed by private business rather than the government.

In the 1960s and 1970s, private initiatives such this seemed unrealistic given the absolute dominance of state-funded space research. But is government-funded space research really the rule and private space research the exception? This perception might hold throughout the second half of the twentieth century, which was marked by NASA's government-funded Apollo program, that led to the moon landing, followed by the space shuttle program, and the ISS. But when Robert Heinlein wrote his book in 1949, there was already a significant tradition of privately funded space research, while primarily government-funded space exploration was still in its infancy. It would be another eight years before the Soviet Union launched the first Sputnik into space, and another nine years before NASA was founded.

Former NASA chief economist Alexander MacDonald's excellent book *The Long Space Age*, published in 2017, deserves recognition for demonstrating that private space exploration played a far greater role than previously thought. Its central finding: "If we look at the history of American space exploration on a longer timescale, a very different history emerges—one in which personal initiative and private funding is the dominant trend and government funding is a recent one. The long-run history thus turns the conventional wisdom on its head: it is the governmental leadership of space exploration that is the more recent phenomenon, while the resurgence of private-sector space efforts in the early twenty-first century represents a return to an earlier pattern."[146]

One example cited by MacDonald is the emergence of ever-larger and increasingly expensive observatories in the United States. It is certainly an unusual approach to include observatories alongside space probes, rockets, and satellites, but fundamentally they served the same purpose: space exploration. Starting in 1830 and lasting about four decades, there was a real "American Observatory Movement."[147] This movement was predominantly funded by private sources rather than the government and was particularly fashionable among the wealthy. For example, almost half (nineteen) of the forty richest men in Boston donated to the Harvard College Observatory.[148]

MacDonald's research shows "that space exploration has mobilized major societal resource commitments in America not for decades but for centuries; that private funding has been a principal driver for the sector; and that funding for space exploration projects equivalent in cost to that of small unmanned spacecraft has been relatively common throughout the Long Space Age."[149]

Of the thirty-eight observatories listed by MacDonald, only two—the U.S. Naval Observatory and the Observatory of the U.S. Military Academy at West Point—were not privately owned.[150] In most cases, it was wealthy entrepreneurs who were interested in space exploration and at the same time signaled their social status through their large investments.

At the end of the nineteenth century, the United States had a higher proportion of superrich individuals relative to the size of its economy than it does today. In 1882, the country was home to over 4,000 millionaires, equivalent to there being more than 4,000 individuals with over $1.1 billion in equivalent 2015 GDP-ratio terms.[151] In fact, there are about 900 billionaires in the United States today.[152]

One notable example of a privately funded observatory from this era is the renowned Lick Observatory, financed by entrepreneur James Lick, the wealthiest man in California at the time. Born on August 25, 1796, in Stumpstown, Pennsylvania, Lick grew up in modest circumstances. He exhibited an early aptitude for craftsmanship and learned the trade of carpentry. He later specialized in the manufacture and, above all, the sale of pianos. In the 1820s, Lick emigrated to South America and lived in Argentina, Chile, and Peru, among other places. He produced high-quality pianos that were very popular with the local upper class, marking the beginning of his financial success.

With his growing wealth, Lick returned to the United States in 1848 and settled in San Francisco, California, during the California

Gold Rush. Rather than searching for gold himself, he recognized the region's economic potential and invested in land and real estate. He acquired property along Market Street, now a central area of the city, which subsequently appreciated significantly in value. In San Jose, he owned over 1,200 hectares of fertile farmland, where he cultivated walnuts, plums, and other fruits. He also held substantial properties at Lake Tahoe and in Napa Valley, often as speculative investments. Particularly noteworthy was his ownership of a substantial portion of Santa Catalina Island, which he envisioned as a resort or for agricultural use. In San Francisco, he commissioned the construction of "Lick House," an ultramodern hotel at the time and considered one of the most luxurious hotels on the West Coast. In the 1860s, his real estate investments made him the wealthiest man in California, and he was known for strategically investing in land close to infrastructure projects.

The real estate investor was somewhat eccentric, at least when it came to his appearance: "Ill-sorted clothes flung carelessly on him, as if time spent in dressing was time wasted."[153] Lick's fortune at the time, $3.5 million, would be worth several billion today.

Lick was interested in space exploration. George Schönewald, the manager of the luxurious Lick House, was walking down Montgomery Street with Lick one day when the businessman suddenly stopped to exclaim: "One day men will go to the moon and back." Schönewald thought Lick crazy, but said nothing.[154]

In 1874, Lick stipulated in his will that a large portion of his fortune should be used to fund charitable projects after his death. The most notable of these was the establishment of an observatory on Mount Hamilton in California. He allocated $700,000 for this project—an exceptionally large sum at the time, which today, measured as an equivalent to U.S. GDP, would be approximately $1.8 billion.[155] In contrast, he left his illegitimate son a mere $3,000.[156]

The Lick Observatory was the first of its kind to be constructed on a mountaintop. Initially, Lick considered building it on Market Street, a property he owned, but friends dissuaded him: "Mr. Lick, I am not an astronomer, but I know well enough, Market Street is not the place for an observatory. In the first place, the climate is too foggy, and, in the second place, the traffic on the streets would disturb the instruments."[157]

Ultimately, Lick was persuaded to build the observatory, which he envisioned as the largest in the world, on Mount Hamilton. The site offered optimal conditions for astronomical research, with approximately 330 clear nights a year, exceptionally stable atmospheric conditions, and no light pollution. Moreover, its elevation placed it above the fog that often envelops the San Francisco Bay area. Since all building materials had to be transported to the site by horse-drawn and mule-drawn carts, a dedicated road was needed.

Construction of the observatory commenced in 1876 and concluded in 1887. It went into operation in January 1888 and was handed over to the University of California in April of the same year. At its inauguration, it had the world's largest refracting telescope, with a 36-inch aperture. "The Lick Observatory, which became one of the central institutions of American astronomy and astrophysics in the decades after its construction, is an example of the significant resources that can be mustered by the private sector for the exploration of space," writes MacDonald.[158] James Lick, who died in 1886, was buried beneath the historic refractor telescope in 1887. A simple brass plaque bears the inscription: "Here lies the body of James Lick."

While personal interest and the desire to create a lasting legacy were central to Lick's motivations, economic motives were more important to other entrepreneurs. One such example is Hulbert

Harrington Warner (1842–1923), an American businessman who amassed great wealth in the late nineteenth century through the sale of patent medicines. At the time, these medicines offered a lucrative market and a way to become very rich, very quickly, with advertising often promising miraculous cures that rarely materialized.[159] Warner was particularly successful with his brand "Warner's Safe Cure," a purportedly curative liquid for kidney and liver ailments. The medicine was sold in eye-catching glass bottles embossed with a safe—a symbol of security and efficacy. Warner marketed the product aggressively and internationally, which made him a fortune.

In 1880, Warner invested $100,000 to build an observatory in Rochester, New York. For him, this was also a marketing tool to promote himself and his medicine. In 1881, he also established the Warner Sage Remedy Prize, which would award a substantial sum of money to any American who discovered a new comet. Initially, he offered $200, later increasing it to $1,000. His observatory was one of the first to be open to the public and soon became a tourist attraction. "The prestige and publicity associated with astronomy could be shrewdly exploited for advertising and commercial gain in the nineteenth century."[160]

Another example is the Mount Lowe Observatory, which was part of a sprawling mountain railway and tourism project led by Thaddeus S. C. Lowe in the late 1890s. Lowe was not only a pioneer of military aerial reconnaissance but also a significant entrepreneur and inventor. He developed the Lowe hydrogen process, which revolutionized the large-scale production of hydrogen for lighting and industrial use. At a time when gas lighting was becoming increasingly widespread, this was an extremely lucrative market.

Lowe secured numerous patents in the field and licensed his processes to gasworks in several U.S. cities, generating considerable income. His innovations also made significant contributions to ice and refrigeration technology, and he founded the Lowe Gas and Ice Company, opening up a new market in the cooling and storage of food. Here, too, he filed several patents and, in 1886, was awarded the prestigious Elliott Cresson Medal for his inventions, one of the most distinguished scientific and technical honors of its time.

During the 1890s, Lowe invested a large portion of his wealth in a visionary infrastructure project, the Mount Lowe Railway in California, an electrically powered mountain railway that combined tourism, technology, and the experience of nature. As part of this railway project, the Mount Lowe Observatory was constructed in 1894, featuring a cutting-edge twelve-inch refractor telescope. This observatory, too, was open to the public, offering visitors a rare opportunity to make astronomical observations and use the instrument under guidance—a very progressive idea for the era, offering both educational value and tourist appeal.

MacDonald writes: "The Mount Lowe Railway was one of the most popular attractions in the Los Angeles area in the late nineteenth and early twentieth centuries and was a forerunner of Disneyland-style family entertainment. The observatory, while only one of the many attractions, was an important draw. It featured prominently in the brochure and was open for public exploration of the heavens three nights a week. Even in the late nineteenth century, individuals within the leisure and entertainment industries understood the value of space exploration in attracting the attention, and the coinage, of the general public."[161]

A further notable example is the Hale Observatory, named after the astronomer George Ellery Hale. He was fascinated by astronomy from a young age, especially after reading Jules Verne's novel *From the Earth to the Moon*.[162] He persuaded his father, a Chicago businessman who had made a fortune in the elevator industry, to give him $25,000 for a private observatory. Throughout his life, he would strive to contact wealthy businessmen who would build other observatories. A diary entry from 1922 reveals that he was writing twenty letters a day at that time, either for fundraising or to cultivate relationships with wealthy entrepreneurs.[163]

Hale's first major investor was Charles Yerkes, a streetcar entrepreneur from Philadelphia, who arrived in Chicago in 1882 and played a pivotal role in the modernization and expansion of the city's public transportation system. After gaining control of two of the three major streetcar companies in the city's north and west, Yerkes converted the horse-drawn tramlines first to cable car operation and then to electric propulsion. By introducing rapid transit, he expanded the network from less than seventy-five miles to 575 miles within a decade. In 1897, he built an elevated tram line around the central business district. His net worth is estimated at $30 million.

Yerkes, unfortunately, had garnered the reputation of being Chicago's most notorious "robber baron." He defrauded shareholders and partners, insulted newspaper editors, shamelessly bribed city officials, and retaliated against dissatisfied customers who raised concerns about subpar service and faulty equipment. As public opinion turned against him, he made an unsuccessful attempt in 1897 to propose legislation that would have prolonged the transit companies' concessions by fifty years.[164] In an effort to salvage his tarnished image, he gave Hale a total of $349,000 to build an observatory.

While Yerkes had no genuine interest in astronomy itself, only in the prestige it could bring, another donor, the famous Andrew Carnegie, was different. Carnegie, a Scottish American industrialist who was regarded as one of the richest men in the world during the nineteenth century, had made his fortune primarily in the steel industry. In 1892, he had founded the Carnegie Steel Company, which was later sold to J.P. Morgan for $480 million, equivalent to approximately 2.1 percent of U.S. gross domestic product at the time—an enormous sum. In today's money, his fortune would be estimated at around $350 billion. He was considered the epitome of the "American Dream," rising from a poor immigrant boy to a multibillionaire.

Later in life, Carnegie devoted most of his money to philanthropic causes. In his book *The Gospel of Wealth*, he described astronomy as one of the "best fields of philanthropy."[165] The Carnegie Endowment, which he established, financed the construction of the giant telescope on Mount Wilson, and it was Carnegie himself who ensured that $1.4 million was invested in the observatory. Hale regularly sent Carnegie spectacular new images captured with the telescope he funded to sustain his enthusiasm.

While Carnegie had a personal interest in astronomy and exerted considerable influence over his foundation to ensure Hale's project was funded, it was a different situation with John D. Rockefeller, who had become the world's richest man through the oil business—with an estimated fortune of up to $400 billion in today's money. Rockefeller had established the Rockefeller Foundation but, unlike his contemporary Carnegie, Rockefeller took a largely hands-off approach in the day-to-day operations of his funding body. However, Hale's vision for a new telescope was so compelling that Rockefeller, as president of the foundation, personally orchestrated the substantial funding required for the project.

The Rockefeller Foundation provided a total of approximately $6 million for the construction of the two-hundred-inch Hale Telescope at the Palomar Observatory. This substantial financial support began in the early 1920s and continued in several stages, with the largest single contribution in 1934. The foundation not only funded the telescope itself but also facilitated the development of scientific infrastructure at the California Institute of Technology, which was responsible for the project. Without this support, the construction of the largest optical telescope of its time would hardly have been possible. Construction was not completed until 1948, and it remained the world's largest observatory until 1976.

From the mid-twentieth century onward, the financing of large observatories changed, and astronomers began to rely principally on government funds. However, there have always been examples of privately funded observatories, such as the W. M. Keck Observatory in Hawaii. The construction of the two large Keck telescopes, which began in 1985, was entirely funded by private donations—over $140 million came from the W. M. Keck Foundation alone.[166]

Observatories like the Lick Observatory and the Palomar Observatory were—according to the former NASA chief economist—roughly as expensive in today's terms as major NASA missions, such as the *New Horizons* mission to Pluto, the MESSENGER probe to Mercury, or the Mars Exploration Rovers.[167]

And it wasn't just observatories that were privately funded; Robert Goddard, the pioneer of American rocket research, also received more money from entrepreneurs and private sources than from the government. Goddard spent his entire life seeking funding for his research, including from the U.S. military. But: "The most significant financial support for Goddard came from private-sector individuals who shared with Goddard a deeply felt intrinsic desire to explore the limits of flight."[168] What is truly outrageous is that the American government, which initially ignored him, was later forced to pay Goddard's widow substantial compensation for patent infringement in acknowledgment of their use of his designs.

But let's explore Goddard's story from the beginning. Alongside Konstantin Tsiolkovsky and Hermann Oberth, Robert Goddard was one of the most important pioneers of rocket research. Konstantin Tsiolkovsky, a Russian educator and self-taught scientist, laid the theoretical groundwork for space travel in the late nineteenth century. Among his many achievements, he formulated the fundamental rocket equation and was already thinking about space stations and interplanetary travel, although his concepts were purely visionary at the time.

In 1923, Hermann Oberth, a German physicist from Transylvania, then part of Hungary and later Romania, published his work *The Rocket into Planetary Space*, in which he explained the physical principles of space travel to a broad audience. He was among the first physicists to scientifically demonstrate that rockets could, in principle, reach space. His work had a great influence on the engineers of his era, including Wernher von Braun.

Wernher von Braun initially developed the V-2 rocket, the first large-scale rocket to reach space, in National Socialist Germany. After World War II, he moved to the United States, where he became a key figure in the American space program. He led the development of the Saturn V rocket, which landed humans on the moon for the first time in 1969.

The American Robert Goddard, often hailed as the father of modern rocketry, was the first to successfully launch a liquid-fueled rocket. He introduced numerous technical innovations, including staged rockets and guidance systems. Through his theoretical work and practical experiments in the 1920s and 1930s, he laid crucial foundations for subsequent space technologies. Unlike Oberth and Tsiolkovsky, who primarily focused on theoretical work, Goddard conducted numerous practical experiments with his rockets. As von Braun said: "In the history of rocketry, Dr. Robert H. Goddard has no peers. He was first. He was ahead of everyone in the design, construction, and launching of liquid-fuel rockets which eventually paved the way into space."[169]

Goddard was a visionary, much like Elon Musk today. At sixteen, he had enthusiastically read H. G. Wells's science fiction novel *The War of the Worlds* and Garrett P. Serviss's unofficial sequel, *Edison's Conquest of Mars*.[170] At the age of twenty-three, Goddard wrote in an essay: "Just as in the sciences we have learned that we are too ignorant safely to pronounce anything impossible, so for the individual, since we cannot know just what are his limitations, we can hardly say with certainty that anything is necessarily within or beyond his grasp. Each must remember that no one can predict to what heights of wealth, fame, or usefulness he may rise until he has honestly endeavored, and he should derive courage from the fact that all sciences have been, at some time, in the same condition as he, and that it has often proved true that the dream of yesterday is the hope of today and the reality of tomorrow."[171]

Then, at thirty-one, he outlined his ideas on spaceflight in an essay he called "The Navigation of Interplanetary Space," in which he proposed that the moon could be a place to manufacture hydrogen and oxygen for travel to other planets. He also wrote about solar energy as a potential energy source for interplanetary travel.[172]

In the essay, he writes: "From an economic point of view, the navigation of interplanetary space must be effected to ensure the continuance of the race; and if we feel that evolution has, through the ages, reached its highest point in man, the continuance of life and progress must be the highest end and aim of humanity, and its cessation the greatest possible calamity."[173] His motivation therefore mirrored that of Elon Musk, as discussed in the previous chapter.

Goddard was a visionary far ahead of his time when he wrote that humanity must learn to move from planet to planet within the solar system, using hydrogen and oxygen as fuel and the metals they would find on the celestial bodies to build further spacecraft.[174]

But he was not only a visionary, he was also a theorist and engineer. In 1919, he published a paper called "Method of Reaching Extreme Altitudes," in which he became the first scientist to mathematically demonstrate that a rocket could overcome Earth's gravity. He also provided the first scientific proof that rockets could fly in a vacuum.[175]

And yet, Goddard faced public ridicule. For example, *The New York Times* wrote in January 1920: "That Professor Goddard, with his 'chair' in Clark College and the countenancing of the Smithsonian Institution, does not know the relation of action and reaction, and of the need to have something better than a vacuum against which to react—to say that would be absurd. Of course, he only seems to lack the knowledge ladled out daily in high schools."[176]

It was not until July 17, 1969, a day after the launch of a manned mission to the Moon, that the newspaper finally issued a retraction in a dry statement (oddly enough, without mentioning the Apollo mission): "Further investigation and experimentation have confirmed the findings of Isaac Newton in the seventeenth century, and it is now definitely established that a rocket can function in a vacuum as well as in an atmosphere." The newspaper added: "*The Times* regrets the error."[177]

During this period, Goddard gained fame as U.S. newspapers sensationalized him as the man who wanted to fly to the moon in a rocket. This led to widespread belief that such a feat was imminent, despite Goddard never making any such claims. More than one hundred volunteers expressed an eagerness to participate in his venture, prompting Goddard to remark that he needed "less volunteering and more solid support."[178]

One newspaper even published an article under the headline "Wants Pay Passenger for Voyage to Moon": "Wanted: A millionaire who is tired of this earth and would like to travel to the moon. There are many who would like to leave the earth, but they haven't the money to pay the fare."[179]

Several millionaires supported Goddard, most notably members of the Guggenheim family, which was among the wealthiest families in the world at the time. The family's wealth stemmed primarily from the mining industry they had established in the latter half of the nineteenth century. Harry Guggenheim, an avid aviation enthusiast, was one day engaged in conversation with his close friend Charles Lindbergh, who had achieved fame for being the first person to fly solo and nonstop from New York to Paris in 1927 aboard the *Spirit of St. Louis*. The two men were discussing how to reach high altitudes. Guggenheim's wife had read about Goddard and his early test flights in the newspapers and brought him to their attention. Lindbergh contacted Goddard and asked him if it was feasible to fly to the moon in a rocket. Goddard confirmed it was, but added that it would require at least $1 million.[180]

Lindbergh, who quickly developed a friendship with Goddard, put him in touch with the Carnegie Foundation and the Guggenheims, who also began to support his work. Over the next few years, Goddard developed a liquid-fueled rocket and conducted 103 static tests and forty-eight flight tests. During this period, the Guggenheim family alone contributed $188,500 to his efforts.[181]

Throughout both world wars, Goddard sought funding from the U.S. military for his projects. However, despite his persistent efforts, their support was not as generous as he had hoped. In today's terms, Goddard received a total of $26 million from the U.S. military, compared to $47.4 million from private sources such as the Guggenheims and the Carnegie Foundation.[182]

After Goddard's death from throat cancer in 1945, his widow and Harry Guggenheim fought for many years to have his achievements recognized. His widow filed a lawsuit against the government of the United States, alleging infringement of his patent rights and the unauthorized use of his inventions. Wernher von Braun was asked for his opinion, and it was not what the government had hoped for: "If his technical opinion translated into a legal one, the federal agencies would be liable to Goddard's successors for damages. To hide that exposure, the bureaucrats classified everything secret and dug in for a fight," explains Goddard's biographer David A. Clary.[183]

Goddard's widow ultimately won the lawsuit against the U.S. government, and *The New York Times Magazine* wrote: "When the Government recently announced a payment of 1,000,000 for infringing patents by the late Dr. Robert H. Goddard, it specifically acknowledged that the Air Force's Atlas and Thor, the Army's Jupiter and Redstone, and the Navy's Vanguard had used what were essentially Goddard rocket engines and Goddard liquid propellants. Tacitly it acknowledged numerous infringements of other Goddard patents by acquiring rights to more than 200 of them, many so basic that every modern rocket, every guided missile, must use them."[184] For Goddard's widow, the payment was above all a restoration of justice, as she had resented the fact, "that her man had been unfairly treated by an ungrateful government that first ignored him and then shamelessly exploited his work."[185]

The American president Lyndon B. Johnson even proclaimed a "Goddard Day" in 1965: "*Whereas* on March 16, 1926, Dr. Robert Hutchings Goddard successfully launched the world's first liquid-fuel rocket at Auburn, Massachusetts; and *Whereas* this achievement, as well as Dr. Goddard's other pioneering achievements in the theory, construction, and testing of rockets, established a foundation for the development of modern rocketry and made possible the exploration of space; and *Whereas* it is appropriate that the great scientific accomplishments of Dr. Goddard should be remembered and that they should be memorialized on the anniversary of his success; and *Whereas* the Congress, by an Act approved March 12, 1965, has designated March 16, 1965, as Goddard Day and has requested the President to issue a proclamation calling upon officials of the Government and the public to participate in ceremonies, meetings, and other activities held in observance of Goddard Day."[186]

In former NASA chief economist MacDonald's judgment, Goddard's program can be seen as part of a continuum of private funding for American space exploration going back more than a century, "Goddard received over the course of his career a level of funding not dissimilar to what might be required for a nontrivial NASA technology development program of the twenty first century. Or to put it into a more historical context, the next privately funded liquid-fuel rocketry program to receive similar funding in the United States would not occur until the first private launch vehicle efforts of the 1980s."[187]

The economist puts modern private space travel in a completely different perspective when he writes: "In the long historical perspective, the trend in the late twentieth and early twenty-first centuries toward increased funding for space exploration projects coming from the private sector—specifically from wealthy individuals such as Paul Allen, Jeff Bezos, and Elon Musk—is understood not as a new emerging phenomenon but rather as the reemergence of a dominant thread in space exploration that dates back to over a hundred years before Sputnik. Incorporating the history of astronomical observatories into the overall narrative of American space history shows that, in fact, it has been private sources that have supplied the resources for the nation's exploration of the solar system and the universe for most of its history to date."[188]

MacDonald draws another important conclusion from his research: It is often claimed that humanity feels a compelling urge to explore and discover the unknown. In reality, it is not humanity as a whole, but a small number of exceptional individuals who experience this primordial desire to conquer outer space—and who therefore devote significant sums to fund space exploration projects.[189]

CHAPTER 4

STARSHIP—A SHIP FOR THE FUTURE OF HUMANITY

Today, if I were to place a bet on which spaceship will be the first to take humans to Mars, I would bet on Starship. Because Starship is a spaceship of superlatives. As space expert Eugen Reichl says: "Very few realize just how revolutionary this spacecraft really is. Starship will dominate space transport for the rest of the twenty-first century. It's huge, yet cheap to build, it blurs the lines between traditional aerospace and shipbuilding, and draws on influences from automotive engineering. It is versatile. It will be built in a wide range of configurations and has the potential to open up the entire solar system to human exploration."[190]

The name "Starship" is somewhat confusing as it collectively refers to both the first stage Super Heavy booster rocket and the Starship spacecraft—the rocket's second stage.

Saturn V, which took the first humans to the moon in 1969, was the largest and most powerful rocket to date. Standing approximately 110 meters tall, Saturn V was only slightly smaller than the latest prototypes of Starship, which measure 124 and 126 meters in the V2 and V3 versions. At the very top, Saturn V had a nearly fourteen-meter-long combination of the Apollo service module, command module, and escape tower.

The actual Starship spacecraft, today increasingly referred to simply as the "Ship," stands fifty-two meters tall, and the entire system, including the boosters, has a launch weight of around 5,000 tons, roughly 1.7 times the weight of Saturn V, which weighed approximately 2,950 tons. According to Elon Musk, future iterations of Starship will be able to carry up to one hundred people to Mars per flight. For such missions, which will take about seven months each way, Starship will, of course, be much more comfortable than the Apollo capsule, which was designed only for eight- to twelve-day lunar missions and had very limited interior space.

Starship's most distinctive feature is also immediately apparent when it is directly compared with Saturn V. Like all rockets of its generation, Saturn V could only be used once, making it very expensive. This is why Elon Musk has spent so long ensuring that his rockets are reusable. He already achieved partial success with his standard launch vehicle, Falcon 9. Starship is designed with both the upper and first-stage booster, known as the Super Heavy, being reusable. The first stage of the rocket returns to Earth shortly after launch, allowing for its reuse in future missions.

Similarly, the upper stage can also return to Earth once its mission is complete, whether that be hours, days, weeks, or months later. However, some versions will never return to Earth again. They will remain—suitably equipped—at their designated destinations: as habitation modules or propellant depots in Earth orbit, on the moon as shuttle vehicles between the lunar surface and lunar orbit, or as permanent bases on Mars, asteroids, or beyond.

The methods used to return Starship and its booster, the Super Heavy, to earth are quite different. The Super Heavy's return is similar to the first stages of the Falcon 9, involving a series of complex, controlled maneuvers. As it approaches the launch site from which it departed six to seven minutes earlier at five to six times the speed of sound, the Super Heavy's engines point in the direction of flight. A few of its thirty-three engines initiate a reentry burn to reduce its velocity to a safe level. Once the booster has reentered the thicker part of the atmosphere, all of its engines shut down, and atmospheric drag acts as a powerful brake, significantly slowing its descent until it reaches its terminal velocity, the speed at which further aerodynamic deceleration via air resistance is no longer possible. The terminal velocity is approximately 195 to 220 meters per second for the Super Heavy. At this point, three engines fire to reduce the residual speed to near zero and only a single engine remains operational. This provides enough propulsion to keep the booster suspended and maneuver it toward the catch tower, where it is caught by a pair of special catching arms, known as "chopsticks."

For the upper stage of Starship, a different landing procedure has been adopted. Approximately forty-five minutes prior to the scheduled landing, while in Earth orbit, Starship executes a "deorbit burn." This involves firing the engine to decrease speed by about 110 meters per second, enabling the spacecraft to reach the upper layers of Earth's atmosphere after half an orbit. At this point, similar to the technique used by the space shuttle, air resistance takes over the entire braking process until aerodynamic drag has slowed it to terminal velocity, which for Starship is approximately 300 to 350 kilometers per hour. Given Starship's significantly higher initial velocity—approximately seven times that of the Super Heavy—the entry angle into Earth's atmosphere (belly flop) must be precisely calculated; otherwise, the vehicle will disintegrate due to aerodynamic heating.

Aerodynamic control surfaces, grid fins made of heat-resistant materials, then steer Starship into close proximity with the launch tower. During this phase, Starship transitions from horizontal, airplane-like flight to a vertical orientation, heading toward the tower, where it is caught by the chopsticks. Currently, SpaceX uses a combined launch and catch tower ("Mechazilla"), though future plans include separate systems to optimize operations and allow for higher launch rates.

The launch and landing process was witnessed by many during the fifth test flight on October 13, 2024. It was an exhilarating experience to see the booster slow down and then be caught by the two recovery arms in a near-hover. "This launch," stated Reichl, "was a milestone in spaceflight history. It was the first time an orbital-class rocket had ever been captured in mid-air."[191]

But Starship boasts numerous other remarkable features, including the immense power of its engines: When all thirty-three engines are running, the total thrust of even the pre-series versions is twice that of the thrust achieved by Saturn V during the Apollo moon mission launches. Elon Musk has also broken new ground with rocket fuel, with Starship's Raptor engines fueled with liquid methane and liquid oxygen.

Musk chose methane because it can be extracted on Mars. This will significantly reduce the amount of fuel that Starship needs to carry as it will not need fuel for both the outbound and return trips. Musk is planning to send an unmanned rocket to Mars, which will generate methane fuel on-site. This fuel will then be used to refuel a subsequent manned rocket for its return journey to Earth. Methane can be synthesized on Mars using the Sabatier process, which combines CO_2 from the Martian atmosphere with hydrogen. As we will see, this fuel mixture is also less expensive than many others that are otherwise used for rockets.

While Starship has a wide range of potential use cases, including for journeys to and from the moon, the entire design is ultimately focused on a single goal, namely transporting large numbers of people to Mars. Musk has consistently emphasized his vision of regular flights to Mars by the mid-twenty-first century, with the ultimate aim of establishing a thriving colony of approximately one million people on the Red Planet. More on that in the following chapter.

Starship is the latest in a series of increasingly powerful rockets developed by Elon Musk. The journey began with the Falcon rockets, primarily designed for commercial satellite launches, resupply missions to the International Space Station (ISS), and other Earth orbit missions. In contrast, Starship is engineered for heavy-lift transport, including Mars colonization, lunar missions such as the Artemis program, and the transportation of passengers or large payloads. It is a more versatile system with a focus on long-term space exploration.

The story of SpaceX's rocket development began with three Falcon 1 launch failures. It is not uncommon for new rockets to experience launch failures, but SpaceX's early days were particularly nerve-racking. The first problem was working out where the rocket should be launched from. SpaceX had chosen Vandenberg Air Force Base in California, near Santa Barbara. However, the site belonged to the Air Force, which imposed a whole host of nonsensical and impractical rules and requirements: "The Air Force and us were such a mismatch," recalls Hans Koenigsmann, who was responsible for rocket launches at SpaceX at the time. "They had some requirements that Elon and I laughed about so hard that we would have to catch our breath."[192] It became clear that SpaceX needed to find a new base of operations after it was revealed that Vandenberg was scheduled to launch a super-secret U.S. spy satellite. In the spring of 2005, shortly before a planned SpaceX launch, the Air Force told them they would have to wait until its spy satellite was safely launched—but they didn't say when that would be.[193]

As an alternative, SpaceX identified a launch site almost 5,000 miles from Los Angeles, on a tiny island called Kwajalein Atoll. The journey to this remote location required several transfers, resulting in a twenty-hour flight from Los Angeles. The first launches in March 2006, a year later in March 2007, and then in August 2009 all failed, and Musk was running out of money, especially as his car company, Tesla, was also in serious trouble at the time. His friends told him he would have to choose: Tesla or SpaceX; he couldn't do both. "For me emotionally, this was like, you got two kids and you're running out of food," he said. "You can give half to each kid, in which case they might both die, or give all the food to one kid and increase the chance that at least one kid survives. I couldn't bring myself to decide that one was going to die, so I decided I had to give my all to save both."[194]

Friends helped him with a cash injection, and the fourth attempt finally succeeded. After a nine-minute flight, Falcon 1 reached orbit, exactly as planned. It was the first rocket from a private company to manage this feat. Musk left the control room and went to the factory, where he was celebrated like a rock star: "There are a lot of people who thought we wouldn't do it—a lot, actually—but as the saying goes, 'the fourth time is the charm,' right? There are only a handful of countries on Earth that have done this. It's normally a country thing, not a company thing."[195]

Musk's biographer Walter Isaacson takes stock: "Falcon 1 had made history as the first privately built rocket to launch from the ground and reach orbit. Musk and his small crew . . . had designed the system from the ground up and done all the construction on its own. Little had been outsourced. And the funding had also been private, largely out of Musk's pocket. SpaceX had contracts to perform missions for NASA and other clients, but they would get paid only if and when they succeeded. There were no subsidies or cost-plus contracts."[196]

Space expert Erik Seedhouse, professor of applied aviation and spaceflight operations and an astronaut trainer himself, writes in his book *SpaceX: Starship to Mars – The First 20 Years*: "In just seven years of existence, SpaceX had positioned itself as a force to be reckoned with, an achievement that was only possible because it was a private company. Because the company had not been encumbered by government red tape, Musk had been able to develop SpaceX as he wanted."[197]

Having invested so much effort and money to develop Falcon 1, it was expected that Musk would now reap the rewards and make money with launch services using his new rocket.[198] But Musk was already thinking much further ahead. Immediately after the successful fourth launch of Falcon 1, he said: "This is just the first step of many. We're going to get Falcon 9 to orbit next year, get the Dragon spacecraft going, and take over from the space shuttle. We're going to do a lot of things, even getting to Mars."[199]

In June 2010, SpaceX conducted the first test flight of the much heavier and larger successor model, Falcon 9. Its success on the first attempt marked a huge accomplishment—two-thirds of the rockets introduced in the previous twenty years had failed on their first launch attempts.[200]

Falcon 9's biggest breakthroughs, however, came on December 22, 2015, and March 30, 2017. December 22, 2015, is a historic date, as that is the day Falcon 9 landed safely at Cape Canaveral. Never before had any nation or company managed to bring an orbital-capable rocket back to earth safely just minutes after launch. Jeff Bezos's company, Blue Origin, had achieved a safe landing with its rocket about a month earlier, but that was a suborbital rocket, incapable of carrying payloads into orbit. It would take another ten years before Bezos managed to achieve what Musk first accomplished in 2015.

And on March 30, 2017, SpaceX achieved another breakthrough by relaunching a previously used Falcon 9 for the first time, achieving the goal of true, i.e., economically viable, reusability. This puts SpaceX in a unique position to this day: As of February 2026, SpaceX had reused one of its Falcon 9 boosters thirty-three times. The goal Musk set for himself in 2010 was a maximum of ten reuses per booster. No other company has ever flown an orbital launch vehicle more than once.

Over the next few years, the Falcon rocket established itself as the world's most important rocket workhorse. On July 13, 2025, Musk congratulated his team: "Congrats SpaceX team on 500 orbital spaceflight missions!" The 500th Falcon flight had taken place the day before, carrying a payload of twenty-six Starlink satellites into orbit from Vandenberg Space Force Base in California. Today, SpaceX is practically unrivaled: Of the 258 rocket launches worldwide in 2024, 134 were conducted by SpaceX. If SpaceX were a country, it would rank first, followed by China with sixty-six successful launches. Without SpaceX, the United States would be in fourth place behind China, Russia, and New Zealand.[201] In 2025, there were 324 global launches. Once again, SpaceX was far ahead of the pack, accounting for 165, followed by China with eighty-eight, and Russia with seventeen.

Falcon's success was due in part to its high reliability and lower costs compared to its competitors. Harry W. Jones of NASA's Ames Research Center notes that during the period of government spaceflight from 1970 to 2010, average launch costs were $18,500 per kilogram: "A major drop in cost occurred in 2010 with the Falcon 9, to $2,700/kg. The Falcon Heavy further reduces the cost to $1,400/kg. The shuttle's launch cost at $57,700/kg was about twenty times that of the Falcon 9 and about forty times that of the Falcon Heavy. The average 1970 to 2000 launch cost of $18,500/kg is reduced by a factor of seven for the Falcon 9 and by factor of thirteen for the Falcon Heavy."[202]

SpaceX currently charges approximately $70 million for a standard Falcon 9 flight, while its own costs are estimated at $15 to $22 million—roughly 20 to 30 percent of the price the company charges its customers. This highlights the true extent of the cost reduction achieved by SpaceX. Of course, much of this reduction can be attributed to reusability, with only 6 percent of Falcon 9 flights in 2025 using new boosters, while some rockets have flown over thirty times.

Starship, according to Musk's plans, is intended to drastically reduce costs further, to $10 per kilogram.[203] While this vision may be overly optimistic, even a figure of a few hundred dollars per kilogram would be sensational. Such cost reduction will also make it possible to send humans to Mars without exposing them to the risks of prolonged exposure to zero gravity or high levels of radiation—a point I will return to in the following chapter. Costs are relevant because it will only be possible to incorporate complex and heavy technical features, such as artificial gravity via rotation and massive shielding against dangerous radiation, with a significant decrease in launch costs.[204]

But how is Musk achieving such large cost reductions? One crucial breakthrough has already been mentioned: reusability. And it's not as if nobody had been working on this before. In 2011, space experts Heribert Kuczera and Peter W. Sacher published a 251-page book on this very topic: *Reusable Space Transportation Systems*. While they emphasized the necessity of reusable rockets, they also noted with some disappointment: "Since the beginning of the 21st century, RLV activities [RLV = Reusable Launch Vehicles, also commonly called hypersonics programs] suffered a dramatic decrease in international interest, not only as a result of the tremendous cost, but also the focus on available technologies and more conservative systems. One driving argument was that the USA had abandoned its hypersonics programs after a follow-on for the space shuttle became more urgent as a result of the planned phasing out of the space shuttle towards 2020. Since then, hypersonics was pursued at a significantly lower level (e.g., at the academic level and smaller research activity level). The major reason can be detected in NASA's official orientation towards expendable launchers and capsule-type vehicles for manned transportation and their application in future missions to the ISS and the Moon. Budgets were therefore primarily given over to these objectives. . . . It is remarkable that the original objective of achieving cost savings by employing reusable systems was never proven. The space shuttle required too much ground operation work (e.g. inspection, maintenance, and repair) between successive launches."[205]

It is interesting to read this assessment by renowned experts from their 2011 perspective today—knowing that just a few years later, SpaceX was able to accomplish precisely what had previously seemed impossible: a cost-effective, reusable rocket. This would likely have been possible even sooner had government bureaucracy not repeatedly imposed lengthy and at times nonsensical approval processes. Musk recounted one of many absurd examples of the Federal Aviation Administration's (FAA) requirements, describing how he was driven almost insane by months of delays to the ultimately successful fifth test flight of Starship: "We had to do this, SpaceX had to do the study to see if Starship would hit a shark. And I'm like, 'It's a big ocean, there's a lot of sharks. It's not impossible, but it's very unlikely.' So, we said, 'OK, fine, we'll do the analysis. And then, well, can you give us the shark data?' They were like, 'No, we can't give you the shark data.'. . . Eventually . . . we got the data and . . . we could say, 'Yeah, the sharks are going to be fine.' But what—they wouldn't let us proceed with the launch until we did this crazy shark study.

Then we thought, 'OK, now we're done.' They said: 'But what about whales?' I'm like, 'When you look at a picture of the Pacific, what percent of that surface do you see as whale? Because I don't see any.' And honestly, if Starship did hit a whale, it's like—that whale had it coming, because the odds are so low. It's like *Final Destination: Whale Edition*. Fate had it in for that whale.

Then they hit us with: 'Well, what if the rocket goes underwater, then explodes, and the whales get hearing damage?' Like, if this is real. . . . Look, if we could make a rocket go underwater and be a submarine, that'd be a feat of physics we couldn't accomplish . . . . This is why I'm feeling the pain of overregulation."[206]

Another example: Before a test flight, SpaceX had to demonstrate that it had taken precautions not to disturb a species of fish that had last been sighted in the Gulf of Mexico more than forty years ago. In another case, seals were captured, fitted with headphones, and then subjected to a sonic boom to determine their reaction.[207]

A lot of time was lost due to all the approval processes, and Musk was particularly affected because, unlike conventional manufacturers who strive for perfection before even thinking about launching a rocket, he factored in the likelihood that many of his launches would "fail." In reality, these "failed" launches were not really failures because SpaceX gained a lot of useful new information and data that helped improve next time. This process is also known as iterative design, and Musk coined his own term for it: "RUD," or "Rapid Unscheduled Disassembly."[208]

Journalists often don't understand Musk's methods. They report, sometimes with a touch of schadenfreude, on the "failure" of a test flight, even if it yielded important insights from the company's perspective. Musk was driven mad by regulatory zeal and risk aversion. "My fucking brain is hurting," he commented, holding his head. "I'm trying to figure out how we get humanity to Mars with all this bullshit." And then he added: "This is how civilizations decline. They quit taking risks. And when they quit taking risks, their arteries harden. Every year there are more referees and fewer doers."[209]

This was likely a major reason why Musk decided to support Donald Trump in the 2024 presidential election. He hoped that Trump could help him eliminate many of the nonsensical regulations that were hindering his space exploration plans. However, the brilliant entrepreneur Musk has little political acumen and even less understanding of human nature; otherwise, he would have known that Trump demands blind obedience from his employees and subordinates. Such deference, however, goes against Musk's character.

That his alliance with Trump could lead to precisely the opposite outcome became clear when Trump threatened to withdraw Musk's NASA contracts and even to deport him.[210] The concern was that Musk's falling-out with Trump could potentially have an adverse impact on SpaceX. The possible repercussions became apparent just a few days after the feud began. In December 2024, Trump had floated Jared Isaacman, a friend and collaborator of Musk's (you will find his name again in chapter 7), as a candidate for NASA leadership, only to withdraw the consideration on May 31, 2025, after a falling-out with Musk. Trump's justification: Isaacman had made donations to the Democrats.[211] While this may be true, publicly available donation records also show that Isaacson financially supported Republican politicians. Therefore, Trump's argument seems contrived. As is so often the case with Trump, Isaacman's appointment was something of a back-and-forth affair, and in December 2026 he was finally named the fifteenth administrator of NASA.

I think it was a misstep for Musk to seek close ties with Trump. Because, based on general experience, friendships are more likely to turn into animosity when they deteriorate (as has happened very often with Trump) instead of developing into a rather distant relationship.

Back to SpaceX and the question of how Musk managed to reduce costs so drastically. Reusability was just one factor. The most crucial factor, and this cannot be highlighted often enough, was that there was actually an economic incentive to cut costs, which, as we have seen, was not the case with the previously dominant cost-plus programs. The oft-maligned concept of "profit maximization" was the economic incentive that, in the cost-plus programs, went in precisely the wrong direction. Now it had a positive impact in terms of driving cost savings, because Elon Musk knew that every dollar he and his team saved would increase his profit.

So, again doing things differently, he initially decided to manufacture about 70 percent of the components he needed in-house, as opposed to sourcing them from external suppliers at significantly higher prices. SpaceX estimates that, "every dollar sent out of the company actually cost between $3 and $5 based on subcontractor overhead and profit."[212]

Musk cuts outlays by scrutinizing every single cost item. When he asked his team why it would cost $2 million to build a pair of cranes to lift the Falcon 9, they showed him all the safety regulations imposed by the Air Force, most of which were obsolete. In this case, Musk was able to convince the military to revise them, reducing the price of the cranes to $300,000.[213]

Sometimes the savings Musk achieved seemed small, but he was also concerned with the fundamentals: educating his employees to question established routines and constantly seek simpler, cheaper solutions. For example, a valve in a rocket would cost thirty times more than a similar valve in a car. Latches that NASA installed in the space station cost $1,500 each. One of Musk's engineers was able to modify a latch used in a bathroom stall and create a locking mechanism that cost $30.[214]

When Musk heard that the air-cooling system for the Falcon 9's payload compartment would cost more than $3 million, he told his employees to buy a $6,000 home air conditioner and modify it.[215] For an engine that normally cost $2 million, Musk told his team to reduce the cost to $200,000.[216] And time and again, to his employees' surprise, they managed to meet Musk's seemingly "impossible" targets.

For Starship's outer shell, Musk made the unconventional choice of stainless steel in place of carbon fiber, which proved not only superior in many ways but also sixty-seven times cheaper.[217] SpaceX also benefited from Musk's ownership of the automaker Tesla, finding simple solutions inspired by the automotive industry that in some cases resulted in parts that were 90 percent cheaper.[218]

Another advantage is the fuel Starship uses, as Robert Zubrin explains: "Critically, Starship's methane/oxygen propellant is the cheapest of all high-performance rocket propellants. Propellant costs are mostly a non-issue for expendable rockets because they are dwarfed by the cost of the hardware lost on each flight. But once rockets become reusable, fuel costs will matter as much as they do for airlines."[219]

At the Cape Canaveral launch site, rockets are prepared for their flight into space. Launch vehicles are first assembled and tested in an assembly hall. They are then transported to the ground system, where further checks and fueling take place. Finally, the rocket is moved to the launchpad, where it awaits the countdown and the actual launch. Traditional companies spent hundreds of millions of dollars, sometimes more than a billion, on all this equipment. Musk approved only $20 million, and once again, money was superseded by creativity.[220]

Musk would make employees sweat if they didn't understand what he meant by the "idiot index," which referred to the ratio of the total cost of a component to the cost of its raw materials. A component with a high idiot index—for example, a component that cost $1,000 although the aluminum that it was made from cost only $100—indicated, to Musk, that it was likely to have a design that was too complex or a manufacturing process that was too inefficient. "If the ratio is too high, you're an idiot," Musk said.[221]

As a result, Musk's Starship program achieved significantly lower costs than NASA's Space Launch System (SLS). Development costs for SLS were about ten times higher than for Starship. Harry Jones of NASA's Ames Research Center stated in 2025 that NASA was anticipating launch costs of $2 billion alone, perhaps even over $4 billion. This would make the costs ($47,500/kg) almost as high as those of the space shuttle. Harvard economists Matthew Weinzierl and Brendan Rousseau, who specialize in spaceflight, reached similar conclusions, SpaceX projected launch costs of $10 million for Starship (cost price, it should be noted, regardless of how much SpaceX charges its customers). "These cost estimates may be optimistic boasts, but even if Starship's costs are much higher, say, $100 million per launch—that would still be forty-two times cheaper on a per-launch basis that what NASA is spending on SLS, even if Starship launches cost one hundred times what SpaceX says—around $1 billion per launch—that would still be one quarter of the cost of an SLS/Orion flight. And as many space commentators have pointed out, NASA would be getting a much more modern product for that money; Starship will be able to refuel in orbit, is more powerful, can carry more cargo, and, of course, doesn't burn up in the atmosphere or end up at the bottom of the ocean after a single use."[222]

NASA's SLS is sometimes derisively referred to as the "Senate Launch System" due to the strong political influence it protects and the backing it ensures for senators in specific states.[223] This nickname reflects criticism that the program's inception was driven less by scientific or technical priorities and more by political lobbying to secure contracts for companies like Boeing and jobs in regions such as Alabama, where key SLS components are manufactured. As with the space shuttle, political involvement has made rocket production slow and expensive.

Aerospace companies have traditionally constructed their rockets using components sourced from an extensive and intricate network of suppliers. For instance, ULA (United Launch Alliance, a 2006 joint venture between Boeing Defense, Space and Security and Lockheed Martin Space Systems) relied on hundreds of subcontractors operating dozens of facilities across the country. As noted by NASA researcher Jones, this was "a political necessity for a government-funded job program."[224] Ultimately, the system became increasingly inefficient as political considerations—where every state sought a share of the program—often overshadowed objective decision-making. In contrast, SpaceX has achieved success partly because it doesn't have to worry so much about all these political sensitivities.

Sooner or later, Starship's innovative design is likely to be copied by various companies and countries worldwide, as is common with groundbreaking, superior products. China is already making attempts in this direction. The first attempts will fail, but eventually, there will be many rocket types that share similarities with Starship. There will certainly be modifications and alterations, some advantageous and others disadvantageous. This will further reduce costs and eventually lead to a market for used Starship models, making spaceflight affordable for many who cannot afford it today.[225]

Elon Musk, however, is thinking on a massive scale, because his goal is not simply to send humans to Mars, but to colonize it. On July 9, 2025, SpaceX submitted plans to the Texas Department of Licensing and Regulation to build a 700,000 sq ft "Gigabay" complex in Starbase (Boca Chica), Texas. This $250 million building is scheduled for completion by December 2026 and is designed to produce 1,000 Starship rocket systems annually—roughly 83 per month.[226] SpaceX itself describes the "Starfactory" or Gigabay concept on its official website as a manufacturing hub capable of building up to 1,000 Starships per year, in order to send enough crew and cargo to build a self-sustaining colony on Mars.[227] On this, Musk commented: "These numbers—while they are insanely high by traditional space standards—are achievable by humans, because they have been achieved in other industries."[228] In support of his claims, he pointed out that his company, Tesla, produces 1.7 million cars every year.

But despite all the successes, there are always setbacks. While I was writing this book, Starship upper stages exploded during their seventh (January 2025), eighth (March 2025), and ninth test flights (May 2025).[229] The tenth test flight on August 26, 2025, while causing some damage, met many of its test objectives and is considered a significant step forward. The eleventh test flight of SpaceX's Starship on October 13, 2025, was a complete success, demonstrating improved maneuverability for future landings at the launch site, and marking the final test of the Version 2 prototypes before transitioning to the more powerful Version 3.

And the great thing is, SpaceX is not alone. On November 13, 2025, Blue Origin's New Glenn rocket launched from Cape Canaveral Space Force Station in Florida, separating its first stage mid-flight and landing it vertically on a floating platform in the Atlantic Ocean. This marked the first successful recovery of the first stage of the New Glenn rocket, following a failed attempt during its maiden flight in January. After SpaceX, Jeff Bezos's Blue Origin became the second company to succeed in landing the first stage of an orbital class launcher propulsively. The New Glenn is a heavy-lift rocket approximately ninety-eight meters tall with a diameter of seven meters and is powered by seven BE-4 meth engines. Depending on the mission profile, it can transport about forty-five tons of payload into low earth orbit. Elon Musk and Jeff Bezos have thus achieved a feat no government-run space agency in the world has managed, further demonstrating the supremacy of space capitalism.

CHAPTER 5

THE COLONIZATION OF MARS—THE DEFINING TASK OF THE CENTURY

In the first chapter, we explored what motivated Elon Musk to make reaching Mars his lifelong ambition. The goal of colonizing our neighboring planet is how Musk intends to transform humanity into a multi-planetary species. His aim is to settle one million people on Mars. To achieve this, he will need to launch 1,000 Starships, each carrying one hundred colonists, to Mars during each launch window, for ten launch windows, which occur approximately once every twenty-six months.[230] Robert Zubrin, founder of the Mars Society and a significant influence on Musk, envisions 50,000 settlers within half a century.[231]

Musk has repeatedly emphasized that no matter how financially successful SpaceX may be, the company will have failed in his eyes if the primary objective—the colonization of Mars—is not achieved.[232]

Even though very few visionaries are currently focusing on the colonization of the Red Planet, a majority of Americans appear to support a mission to Mars. According to a YouGov poll from June 2025, 65 percent of respondents endorsed the idea of the United States sending astronauts to Mars.[233] However, it should be noted that a Pew Research survey from late May 2023 indicated that other space projects are more important to U.S. adults: When asked what NASA's top priorities should be, Americans ranked monitoring asteroids and other objects that could hit Earth first (60 percent) and sending human astronauts to explore Mars last (11 percent).[234]

Why does Mars attract such particular interest from Musk and other visionaries who want to transform humanity into a multi-planetary species? The other planets and moons in our solar system are either far too distant, far too hot (Venus: 860 degrees Fahrenheit!), or far too cold, and, based on what we know right now, do not offer conditions as conducive to colonization as those of Mars. But what makes Mars more suitable for colonization than other celestial bodies in our solar system, and what challenges still need to be overcome?

Unmanned missions to Mars, including NASA's *Curiosity* rover, which deployed a car-sized, remote-controlled robot-laboratory on the planet in 2012, and *Perseverance*, which has been investigating signs of ancient microbial life on Mars since 2021, have provided us with a detailed picture of the planet. Water is a fundamental requirement for human life and agricultural endeavors. Due to the extremely low temperatures and atmospheric pressure, water does not exist in liquid form on the Martian surface. However, it is present in frozen form. Substantial quantities of water ice have been discovered at the polar ice caps and in higher latitudes by various missions, including the European Space Agency's *Mars Express*.

In July 2018, the first reports of a possible subsurface saltwater lake on Mars emerged. Roberto Orosei and his team published their findings, "Radar evidence of subglacial liquid water on Mars," in the journal *Science*. Using a radar instrument aboard *Mars Express*, they detected anomalously bright reflections about one mile below the south pole ice cap, indicating a large body of water, about twelve miles wide, matching the characteristic signatures for liquid, saline water.[235] Further investigations in 2020 identified three additional smaller underground basins near the main lake, which is now estimated to be approximately nineteen miles wide. Other ground-penetrating radar observations have shown the existence of massive glacier formations composed of pure water ice covered by just a few meters of dust, covering much of the planet's northern region, spreading southward from the pole as far as 38 degrees North—the latitude of Athens on Earth.

Frozen water on Mars could be extracted, melted, and purified to produce drinking water, oxygen (via electrolysis), or even hydrogen fuel. Experiments conducted by NASA and other research institutions show that plants can be cultivated under Mars-like conditions in special greenhouses. The Martian soil—consisting of regolith—contains many essential minerals for plant growth, such as phosphorus, potassium, and magnesium. However, the presence of perchlorates, toxic salts detrimental to both plants and humans, poses a significant challenge. These would need to be removed through chemical or biological processes before the soil could be used. Alternatively, hydroponic systems could be used, in which plants grow in a nutrient solution, without "soil" or other loose material.

Since the natural environmental conditions on Mars—low temperatures, low air pressure, and strong UV and cosmic radiation—make it impossible to grow plants in the open, colonists would need controlled greenhouse environments with regulated temperatures and humidity. These could use natural sunlight if positioned on the Martian surface. If they were underground they would need artificial lighting.

Alongside growing crops, energy supply is one of the biggest challenges a Mars colony would face. Solar radiation on Mars is only 43 percent of that on Earth, which significantly limits the efficiency of solar power plants. Furthermore, massive dust storms, which can last for weeks or even months, would obscure the solar panels and severely impact energy production. While solar energy could serve as a supplementary energy source, compact nuclear reactors are considered a solution for providing a reliable and continuous energy supply. Reactors based on technologies like those being developed by NASA as part of the Kilopower project could be transported to Mars aboard spacecraft like Starship. These reactors offer high energy density and are ideally suited to the harsh Martian environment, operating independently of weather conditions or time of day.

In the long term, other technologies, such as geothermal energy—assuming viable geothermal sources can be found on Mars—could also be considered. Besides the availability of water and its potential agricultural applications, Mars offers additional advantages over other celestial bodies in our solar system: The length of a day on Mars is 24.6 hours, very similar to a day on Earth, making adaptation easier for humans. Mars also has a thin atmosphere composed mainly of carbon dioxide, which could be used for both plant cultivation and fuel manufacturing, for example, by converting it into methane via the Sabatier reaction. As previously mentioned, this is the main reason Musk decided to use methane as fuel for Starship.

Given the absence of oxygen and the low atmospheric pressure on Mars, habitats would need to be airtight and pressure-sealed to maintain Earth-like conditions. Shielding from cosmic radiation would also be required. Initial settlements would, therefore, most likely be established with their sleeping quarters underground, potentially covered by large domes to allow sunlight to pass through to illuminate areas suitable for outdoor activities. The cosmic ray dose rate on Mars is approximately the same as that at the International Space Station. By limiting aboveground activities to a fraction of the day, doses could thus be kept well within safe limits.[236]

Mars, much like the Moon, is likely to have large lava tubes—tubular subterranean cavities formed by solidified lava. These structures develop during volcanic eruptions when the outer layer of lava cools and solidifies while the interior layer continues to flow, leaving a hollow space. Mars is home to numerous large volcanoes (e.g., Olympus Mons, the largest probably inactive volcano in the solar system), and there is ample evidence of ancient lava flows. Satellite images, including data from the Mars Reconnaissance Orbiter, have revealed features known as skylights—openings that likely lead into underground lava tubes. Some of these Martian lava tubes could be ideal locations for future human settlements, offering natural protection from radiation, micrometeorites, and extreme temperature fluctuations.

Even today, some countries have colossal, self-contained shopping malls that provide an idea of what a city on Mars might look like. For instance, the Dubai Mall spans over 12 million square feet and includes an aquarium, an ice rink, numerous restaurants, and hotels, all within a fully air-conditioned environment. Or the West Edmonton Mall in Canada, also a completely self-contained structure, which, in addition to shops, also features extensive leisure facilities, including a water park, a hotel, and an ice rink.

The comparison between life on Mars and life in a shopping mall is often drawn, including in the book *Terraforming Mars*: "People going around, eating and enjoying themselves in an air conditioned shopping center located in any city in very hot or cold countries, do not live in a less artificial environment than future Mars colonists living in their pressurized dwellings. . . . The feeling of living in an environment separated from the outside is similar."[237] However, constructing such dwellings to be pressure-tight on other celestial bodies presents a completely different technical challenge.

There are many ideas about how humans could live on Mars. Again, in the book *Terraforming Mars*, Chris Hajduk writes: "This Martian colony design is based on the principle of extracting water out of the soil and then redeveloping the holes that had been dug during the water extraction and turning them into apartments and buildings with a central 'public park' looking out onto the Martian sky and bringing in natural sunlight into each apartment and building without direct long-term exposure to radiation."[238]

In his books, Robert Zubrin establishes that colonizing and building cities on Mars can be done. But who will pay for it? He suggests that the first Mars missions would perhaps be state-funded, while the long-term goal of establishing a colony on the planet would require private-sector innovation and investment: "While a Mars base of even a few hundred people can probably be supported out of pocket by government expenditures, a developing Martian society, one that may come to number in the hundreds of thousands, clearly cannot. To be viable, a real Martian civilization must be either completely autarkic (very unlikely until the far future) or be able to produce some kind of export that allows it to pay for the imports it requires. Around this question will hang the future of Mars, and not just human civilization on Mars but the very nature of the planet itself. If a viable Martian civilization can be established, its population and powers to change the planet will continue to grow. Mars was once a temperate planet, and with enough work, it can be made so again. The advantages to Mars settlers of a terraformed world are so obvious, that put simply, if Mars is colonized, then it will also be terraformed."[239]

The Mars colony, Zubrin stresses, would enjoy unique advantages, such as the fact that it is much easier and more cost-effective to access the resource-rich asteroids between Mars and Jupiter from Mars: "There will be a 'triangle trade' with Earth supplying high-technology manufactured goods to Mars, Mars supplying low-technology manufactured goods and food staples to the asteroid belt and possibly to the Moon as well, and the asteroids sending metals (and perhaps the Moon sending helium-3) back to Earth."[240]

Zubrin believes that the challenging living conditions on Mars and the constant need to find novel solutions to problems would make a Mars settlement a terrific engine of invention. This would create IP for export, ultimately greatly benefiting the Earth. "In short, Martian civilization will be practical because it will have to be, just as nineteenth century American civilization was. This forced pragmatism will give Mars an enormous advantage in competing with the less-stressed and therefore more tradition-bound society remaining behind on Earth. If necessity is the mother of invention, Mars will provide the cradle."[241]

Zubrin has faith in the inventiveness of the colonists and believes that they would be able to establish a thriving economy and society on Mars, much like the early European settlers who paved the way for the United States to become the most successful nation in history.

Despite all of the above, there are many skeptics who explain to us that colonizing Mars would not be economically viable. Matthew Weinzierl and Brendan Rosseau believe that "any economic return would be highly uncertain, as exports from Mars—if they were to exist—would be extremely costly to transport, and demand among wealthy Earthlings to live there is likely to be quite limited, at least at first."[242]

That is correct from today's perspective. However, if the cost of space travel continues to fall as dramatically as it has over the past ten years, then trips to Mars will no longer be affordable only to the very wealthy. Perhaps it will one day be no more expensive than migrating from Europe to the United States used to be, and someone who sells their house on Earth will be able to finance their flight with the proceeds. Musk and Zubrin estimate that the cost of a trip to Mars could be as low as $200,000 or $300,000,[243] which is significantly below the average price of a house in the United States.

And as for exports, it's wrong to think primarily of heavy goods. Zubrin's speculations mostly focus on patents, because the inhospitable and unfamiliar conditions on Mars will inevitably drive innovation, particularly in areas such as food production, nuclear fission and fusion, geothermal energy, robotics, and AI: "In my view, the best, early, large-scale source of cash income that Mars colonists can generate will come from the sale and licensing of intellectual property. This will come naturally from the nature of the Martians themselves and their situation. The Martians will be a group of technically adept people in a frontier environment that will challenge them, indeed force them, to innovate. They will face a terrific labor shortage. This will compel them to innovate in the areas of labor-saving machinery, automation, robotics, and artificial intelligence. Limited to greenhouse agriculture, they will have a shortage of land and livestock. This will force them to innovate in the area of biotechnology, to create ultra-productive and highly nutritious crops. Lacking attractive sources of fossil fuels, wind or water power, or solar energy, they will be impelled to innovate in areas of nuclear power, including advanced fission designs . . ., and fusion, as the deuterium fuel for fusion reactors is five times as common on the Red Planet as it on Earth. All these innovations will have tremendous utility on Earth. The Martians therefore will patent them and license the patents for use on the home planet. The revenue from such intellectual property sales could be enormous."[244]

I think Zubrin is right and Martian cities will look very different from cities on Earth. For one thing, robots will perform absolutely every task they can. Unlike humans, robots will not be significantly affected by the pressure, lack of oxygen, or radiation of the inhospitable Martian environment. They can work twenty-four hours a day without food, breaks, or vacations. Which is why the small Martian settlements in my science fiction novel *2075* contain considerably more robots than humans.[245]

There is also potential for profitable real estate transactions on Mars. I will discuss the topic of property ownership later, in chapter 10. But, assuming private property exists on Mars, and colonization would be impossible otherwise, a lively market for plots of land could quickly develop, attracting investors from Earth who may speculate on the value of Martian land in the hope of reselling it later at a higher price, particularly plots close to potential water sources. Mars would also be interesting for the advertising and entertainment industries: Its gravity, which is only 38 percent of Earth's, opens up possibilities for novel sporting competitions to be held there.

Some authors believe that Mars will attract tourists, including climbers keen on conquering Mars's Olympus Mons, which is about 2.5 times taller than Mount Everest (and so wide that, if placed on North America, it would stretch from New York City to Montreal,

Canada).[246] In his book *The Future of Humanity*, the physicist Michio Kaku enthuses: "Walking across the Martian terrain would be a hiker's dream."[247]

Indeed, given the breathtaking images that probes have sent us from Mars, it's easy to imagine. But I consider large-scale tourism on Mars unlikely, since not only does it take six to nine months to reach the planet, there's also only one launch window every twenty-six months. A tourist would need to commit several years of their life to such a journey, and that's something only a few wealthy retirees would realistically be able to do. Tourists are more likely to travel to the Moon, which would be worth a three-week trip.

Society on Mars would also have to be organized differently than on Earth. For one thing, along much more rational lines. A Martian society would necessarily tend toward a minimal state, similar to the early United States, because tolerating unproductive individuals at the expense of the productive would be unsustainable.

The long-term vision of Zubrin, Musk, and other visionaries involves "terraforming" Mars. The term "terraforming" was first used in 1930 by Olaf Stapledon in his novel *Last and First Men*, in which he describes how, with Earth no longer habitable, humans terraform Venus. The term "terraforming" became more widely known thanks to the science fiction author Jack Williamson, who used it in his 1942 short story "Collison Orbit," before it was later adopted by the scientific community at large.[248]

Terraforming refers to the use of technologies to transform hostile planets into habitable ones over several centuries. Proposed technologies include gradually warming the Martian atmosphere, essentially creating an artificial greenhouse effect, so that liquid water can flow again on Mars for the first time in three billion years. Physicist Michio Kaku writes: "To initiate the process of terraforming, we might inject methane and water vapor into the atmosphere to induce an artificial greenhouse effect. These greenhouse gases would capture sunlight and steadily raise the temperature of the ice caps. As the ice caps melt, they would release trapped water vapor and carbon dioxide."[249] Zubrin and NASA scientist Christopher McKay propose using fluorocarbon gases instead, as they are much stronger greenhousing agents than methane and, unlike methane, are resistant to destruction by solar ultraviolet.[250]

Another idea is to orbit Mars with satellites equipped with gigantic sheets, many miles across, containing a vast array of mirrors or solar panels. Sunlight could either be focused and then aimed toward the ice caps, or the energy could be converted using solar cells and then sent down as microwaves.[251]

Then there are even more radical ideas, such as "to change the orbits of small asteroids or comets and drop them onto Mars. . . . The kinetic energy of the asteroid or comet would be converted entirely into thermal energy, causing localized heating around the point of impact and releasing the materials from which it was made."[252] Giancarlo Genta concedes this might sound crazy, "but this mimics the natural mechanism of planetary foundation. It is thought that, during its accretion, the Earth received water and carbon (the latter perhaps already having formed organic compounds) from falling cometary nuclei."[253]

This idea sounds as interesting as it does risky, because by that time humans would already be living on Mars, and the process would have to be controlled in such a way as to avoid posing a risk to them.

A completely novel idea was published by a team of scientists led by Professor Edwin Kite of the University of Chicago in August 2024 and would lead to success in a more environmentally friendly way. They propose warming Mars using nanoparticles: "One-third of Mars's surface has shallow-buried H2O, but it is currently too cold for use by life. Proposals to warm Mars using greenhouse gases require a large mass of ingredients that are rare on Mars's surface. However, we show here that artificial aerosols made from materials that are readily available at Mars—for example, conductive nanorods that are ~9 micrometers long—could warm Mars >5 × 10^3 [= 5,000, R.Z.] times more effectively than the best gases. Such nanoparticles forward-scatter sunlight and efficiently block upwelling thermal infrared. Like the natural dust of Mars, they are swept high into Mars's atmosphere, allowing delivery from the near-surface. For a ten-year particle lifetime, two climate models indicate that sustained release at thirty liters per second would globally warm Mars by $\gtrsim$30 kelvin and start to melt the ice. Therefore, if nanoparticles can be made at scale on (or delivered to) Mars, then the barrier to warming of Mars appears to be less high than previously thought."[254]

Terraforming is by no means an "all or nothing" situation, as expert Martin Beech explains in his books. Even if the ultimate goal of transforming Mars to allow humans to move as freely on our neighboring planet as they do on Earth remains elusive over the next few centuries, incremental steps (such as making it possible for certain crops to grow outside of greenhouses) would represent significant progress.[255]

But many questions remain unanswered. Michio Kaku writes that if the Martian ice caps were completely melted, there would be enough liquid water to fill a planetary ocean fifteen to thirty feet deep. But would the changes induced by terraforming, if successful, be permanent, or would an artificially generated magnetic field need to be created around Mars?[256] Undoubtedly, a whole host of completely different problems will emerge that we cannot even imagine today. At the same time, technological advancements and the ingenuity of creative entrepreneurs will also yield solutions that we currently consider impossible.

But that's all speculation and won't happen this century anyway. So, let's return to the present: First and foremost, the first manned missions to Mars are on the horizon. Even more than fifty years after the moon landing, NASA hasn't managed to send astronauts to Mars, nor has any other nation. On July 20, 1989, U.S. President George H. W. Bush announced his "Space Exploration Initiative": "First, for the coming decade—for the 1990s—Space Station Freedom. . . . And next—for the new century—back to the Moon. . . . And then—a journey into tomorrow—a journey to another planet—a manned mission to Mars."[257]

The topic of Mars had been in the air for some time, and a movement originally spearheaded by young American scientists, calling itself "The Mars Underground," was holding conferences as long ago as the early 1980s to demonstrate the feasibility of a Mars landing.[258] The Mars movement was thus initially a spontaneous movement, emerging from the "bottom up," so to say.

Three months after Bush's speech, NASA unveiled a study called "Report of the 90 Day Study on Human Exploration of the Moon and Mars." It was an ambitious program, calling for a vast space station containing large hangars for the construction of interplanetary spacecraft, which would then transport materials to the Moon, where a massive lunar base complex would be built. This lunar base, would be used to build "Battlestar Galactica-class" spaceships weighing more than a thousand tons, which would then, finally, transport humans to Mars.[259] The costs were so immense that the report's authors didn't even dare to include them in the plan. However, cost estimates for the program leaked to the press, amounting to about $450 billion (roughly $1.1 trillion today).[260]

Robert Zubrin, who was then working for the aerospace company Martin Marietta, immediately recognized that this plan was completely impractical and far too complicated and expensive. The U.S. Congress, he knew, would never approve such a plan. Moreover, he explained, it was important to remember that: "Every year any major program has to go before Congress for continued funding where it faces risk of termination, often caused by deals or interpersonal frictions that have nothing to do with the program itself. Every time a program goes before Congress for funds it is forced to play another game of Russian roulette."[261]

This was also Zubrin's argument for creating a ten-year plan for the Mars mission instead of a thirty-year one. Here is his critique, which he presented at the time in response to the NASA report: "The plan, in order to be simple, robust, and low cost, should not make inter-dependent missions (i.e. Lunar, Mars, and Earth orbital) that have no real need to be dependent on each other."[262]

Trying to achieve all of these ambitious missions at once would, Zubrin warned, ultimately mean failing to achieve any of them. Furthermore, he pointed out, the basic question of why humanity even wants to go to Mars remains unanswered in NASA's plans. "It is not enough to go to Mars, it is necessary to be able to do something useful when you get there. Zero capability missions have no value."[263]

It made no sense, Zubrin warned, to rely on a low earth orbit infrastructure, such as a large space station, as a prerequisite for missions to the Moon or Mars. And for cost reasons alone, he advised that the same rockets and propellants should be used for both the lunar and Mars missions. Additionally, Zubrin explained, NASA really needed to develop a strategy as to how the fuel required to return to Earth could be produced on Mars itself. "We thus see that the requirements for Simplicity, Robustness, Low Cost and High Effectiveness drive SEI toward and architecture utilizing direct launch to the Moon or Mars with a common launch and space transfer system, and direct return to Earth from the planet's surface utilizing indigenous propellants, which are also used to provide surface mobility and mobile power."[264]

Zubrin and his colleague Ben Clark, another employee at Martin Marietta at the time, argued that the company should develop an alternative to the NASA plan. This was a radical idea, especially as aerospace companies such as Martin Marietta had always subscribed to the industry wisdom that proposals had to be formulated in such a way that the "customer" (in this case NASA) would like them. Typically, this meant that every stage of a proposal, from the original concept onward, was not designed with the aim of developing the best possible plan to put people on Mars and then trying to convince NASA of the plan's merits at a later date, but rather with the aim of pleasing government decision-makers and NASA managers by formulating the plan in such a way that it would align with their existing approaches and gain their approval. "We were proposing the opposite approach: Come up with some good ideas and then tell the customer what they need to hear, whether they like or not."[265]

Zubrin and his colleagues crafted a detailed plan based on these principles, which was designed to reduce costs by eliminating the need for a space station or moon base, among other things. One key aspect of their plan involved transporting the equipment to make the fuel for the return trip from Mars to Earth out of local materials in a separate spacecraft. In the spring of 1990, Zubrin presented his "Mars Direct" proposal to various NASA departments and the reaction was far more positive than he had ever dared hope. In fact, he was met with unequivocal enthusiasm. On Memorial Day, May 27, 1990, Zubrin presented the Mars Direct plan publicly for the first time at a National Space Society event in Anaheim. He received a standing ovation, and his plan was also reported in the press for the first time.[266]

However, the plan soon encountered resistance. Powerful forces within NASA linked to the space station program viewed Zubrin's proposal, in which he argued that a space station was unnecessary for a Mars mission, as a direct threat. Zubrin's colleague, David Baker, a key supporter of the Mars Direct plan, became so disillusioned with NASA's obtuseness that he quit and went back to university. But Zubrin did not want to give up and convinced NASA to embrace a design reference mission based on Mars Direct ideas. The modified proposal came with a price tag of approximately $50 billion, double the cost of Zubrin's initial estimate of $20 to $30 billion,[267] which was still only one-eighth of the cost of the original NASA program. *Newsweek* magazine even featured the story on its cover under the headline "A manned mission to Mars?" The article explained: "The technology is already in place. And at $50 billion—one-tenth of previous estimates—it's a bargain."[268]

Zubrin persisted and in 1996 he published a widely acclaimed book, *The Case for Mars*, describing in detail how a Mars mission and subsequent colonization of the Red Planet could be achieved. Drawing on the findings of several unmanned Mars missions, Zubrin concluded that no planet—and no moon—within our solar system was as suitable for colonization as Mars.

However, just getting to Mars presents challenges. Some skeptics raise concerns about the potential dangers of such a lengthy voyage, citing issues such as prolonged weightlessness and, most significantly, radiation exposure. It is true that beyond Earth's magnetic field, astronauts are exposed to radiation from Galactic Cosmic Rays (GCR) and Solar Particle Events (SPE). According to "Space Radiation and Astronaut Health," a study commissioned by NASA in 2021, astronauts on a two-year interplanetary mission would be expected to absorb more than 600 mSv (millisieverts) of radiation—a unit of measurement for the effective dose of ionizing radiation to the human body. This would increase their risk of cancer at some point later in their lives by three percentage points.[269] But since the lifetime risk of developing cancer in the United States is 39 percent,[270] this would mean that a flight to Mars would increase it to 42 percent. While this is significant, compared to the other risks associated with a mission to Mars, it's certainly not a reason to oppose the trip.

Extensive research has already been conducted on the potential use of genetic engineering to adapt humans to the challenges of space. For example, discussions have explored how gene therapy could mitigate the negative effects of prolonged periods in space, such as muscle and bone loss and radiation exposure. For anyone interested in finding out more, I can highly recommend the volume *Human Enhancement for Space Missions: Lunar Martian, and Future Missions to Outer Planets*, edited by Konrad Szocik, which contains nineteen scientific articles that delve into the subject in detail.

What might otherwise be contentiously discussed under the banner of "human enhancement" could perhaps simply be viewed as a necessary and sensible adaptation to the conditions of space: "When it comes to space exploration," writes former Stanford scientist Arvin M. Gouw in one article, "biomedical interventions that would normally have been considered as enhancement would be considered a medical necessity for astronauts. For example, being genetically modified to double muscle mass would be considered to be an enhancement for an athlete, but it would be considered to be a therapy for an astronaut due to microgravity."[271]

Regardless of the risks involved, the colonization of Mars—or any other celestial body—is always going to be controversial for many reasons. The debate centers not only on whether colonization is possible, but also on whether it is ethically justifiable. I will present this discussion in considerable detail in the following because I believe that proponents of space exploration massively underestimate the dangers posed by ideologically motivated naysayers.[272]

In 2023, Kelly and Zach Weinersmith published a book titled *A City on Mars: Can We Settle Space, Should We Settle Space, and Have We Really Thought This Through?* For the Weinersmiths, concerns about colonizing Mars or other celestial bodies outweigh any potential advantages. They argue that while we might be able to think about space settlement sometime in the distant future, we are nowhere near that point yet.

Of the concerns they raise, some are worth considering, including unresolved questions as to whether humans can reproduce safely in low-gravity conditions and whether babies can develop normally in such environments.[273] They do write, however, that the Chinese are already planning experiments in which monkeys will mate on their space station—so perhaps we will get our first answers soon.

Many of the questions in the book, though, are far-fetched and contrived, and it is clear that the authors have adopted the common approach of intellectual worrywarts, who first want a definitive and foolproof plan that answers all conceivable questions before they will even begin to act. This stands in stark contrast to the mindset of the entrepreneur, who takes action and continuously solves new challenges as they arise.

Some of their fears are absurd, for example the idea that states might bombard each other with asteroids.[274] Some of what they write is contradictory, such as when they present the idea that space settlement may not yield discernible economic benefits,[275] while also portraying a grim scenario of nuclear conflicts arising over a scramble for territory on celestial bodies like the Moon or Mars. These and other absurdities were underlined by Zubrin in a devastating review of their book.[276]

The authors do, however, make some valid points: It is indeed unreasonable to believe that an environmental calamity will imminently annihilate all human life, requiring us to find a replacement planet. Even if this threat existed, they argue, it would be too late anyway.[277] And they are also right to reject the idea that colonizing other celestial bodies would eliminate wars on Earth or lead to the creation of a perfect, utopian society.

But many of the points they raise create the impression that they simply wanted to assemble every conceivable reason why we shouldn't go to Mars. For example, they ask how surgery could safely be performed under microgravity conditions; whether couples would have to tether themselves to each other during sex in low gravity; how psychiatric care could be provided for people suffering from mental illness on Mars; whether a permanent stay on Mars would have a negative effect on the psyche, whether medication for mental disorders would be negatively affected by space radiation, and how the profits from asteroid mining could be distributed "fairly."

Countering the claim that "space exploration is a natural human urge," the Weinersmiths argue: "Most of us are not in fact famous explorers. Most of us prefer to vacation in places that have pastries and air-conditioning, not Mount Everest or the Amazon basin. . . . If exploration is a natural human urge that must be satisfied, why are so many of us happy to sit on our couches?"[278]

Ian Stoner makes a similar argument when he writes in the book *Terraforming Mars*: "It is far from clear that the drive to expand is part of human nature. Some people seem driven in this way, but many more are content to stay home."[279] To this I object: Human progress, in whatever field, has never been achieved by those who prefer to stay at home and lie on their couches, but by those who do not accept an average existence, who are different from the majority, who are more curious and perhaps also more adventurous than the masses.

An entrepreneur who, before setting up a business, thoroughly considered every single one of the hundreds of potential problems that could occur, as the Weinersmiths do when they raise myriad concerns against the colonization of space, would end up doing exactly what the authors recommend: "We don't think this has to mean that space settlements should never happen. What we do think is that space settlements probably are, and ought to be, a project of centuries, not decades. . . . We should take a 'wait-and-go-big' approach. Wait for big developments in science, technology, and international law, then move many settlers at once."[280] They would dawdle and delay—and never do anything at all. Just imagine: If the Weinersmiths had been around when our first human ancestors learnt to make fire, they would have said: "Hold on! We need to wait until we've got a perfectly functioning fire brigade and precise plans on what to do if a fire gets out of control. You should wait until we've got every eventuality covered."

In her 2022 book *Astrotopia: The Dangerous Religion of the Corporate Space Race*, the American religious studies scholar Mary-Jane Rubenstein, who is better known for her work on environmental and gender issues, levels an even more fundamental critique of the ambitious plans of Musk and others. Rubenstein takes aim at what she terms "Western antimineralism," which she sees as "a tendency to value those rocks that have been removed, installed, carved, stacked, and shaped by human hands (and market forces) over those rocks that remain where and as geological (and ancestral) processes made them."[281]

This sounds absurd and can only be understood in the context of an anti-Western, "postcolonial" philosophy. The proposals from Elon Musk and others to "colonize" Mars were bound to provoke a backlash from today's modern "postcolonialist" ideologists. Obviously, however, the key distinction between colonizing countries on Earth (such as the Americas) and space is that the Earth is populated with humans, whereas Mars, the Moon, and asteroids are either devoid of life or at best inhabited by microbes. But the postcolonialists do not accept this argument. Rubenstein approvingly quotes the American astrophysicist Carl Sagan, who argued: "If there is life on Mars, I believe we should do nothing with Mars. Mars then belongs to the Martians, even if the Martians are only microbes."[282] In response, Zubrin notes: "The antihumanist's claim saying that harming Martian rocks or microbes is the same as harming Native Americans is extremely racist. Harming Native Americans was wrong because they were humans, not microbes! By stripping humans of their special dignity, the antihumanists are endorsing the most racist positions imaginable."[283] Well, racist might not be the most appropriate term, but the postcolonialists' arguments are certainly misanthropic.

One argument that is probably more popular than insisting on the rights of microbial life-forms is playing off the significant investment in space exploration against efforts to solve other problems. Rubenstein not only opposes the colonization of Mars, she is also against the moon landing, for example. "Neil Armstrong's Moonwalk," Rubenstein declares, "has done very little for poor,

Black, Indigenous, and immigrant people in his own country."[284] And she poses the following question: "How would the late-sixties lunar landing advance the civil rights of Black Americans back on Earth? How would it contribute to resolving the mess Apollo was escaping in Vietnam? How would the moon shot help decolonize India and Africa? What was its stance on the labor movement, the women's movement, reproductive justice, gay rights, food shortages, poverty, dictatorial regimes, refugee resettlement, nuclear proliferation, water rights, and the growing sense that there was something very wrong with the climate?"[285]

These questions are clearly irrelevant, as no one has ever suggested that the moon landing should play any kind of role in the decolonization of Africa. That was never the aim of the mission, nor could it have been. One could just as easily ask what the construction of any major bridge, hospital, or university contributes to the decolonization of India or how it improves the rights of black Americans.

Deondre Smiles, Professor at the University of Victoria, Canada, whose research focuses on "Critical indigenous geographies; human-environmental interaction; political ecology; tribal cultural resource preservation" argued in an essay on "The Settler Logics of (Outer) Space" in 2020 that the perspective of indigenous peoples is given too little consideration in the discussion about space travel: "In one example, when asked about the Moon landings, several Inuit said: 'We didn't know this was the first time you white people had been to the moon. Our shamans have been going for years. They go all the time. . . . We do go to visit the moon and moon people all the time. The issue is not whether we go visit our relatives, but how we treat them and their homeland when we go."[286]

What is Smiles trying to tell us? That the Western perspective on space travel is characterized by one-sided, rationalist thinking and must be expanded to include indigenous perspectives: "These previous examples should serve as a reminder that the historical underpinnings of our great national myth are built upon shaky intellectual ground—we need to be honest about this."[287] Above all, he wants to question "the prioritization of 'science' over Indigenous epistemologies."[288]

These questions have now even found their way into official documents from institutions that advise NASA. A paper published in 2020 by the Equity, Diversity, and Inclusion Working Group of the Planetary Science and Astrobiology Decadal Survey states: "The Moon and other planetary bodies are sacred to some cultures. Is it possible for those beliefs to be respected if we engage in resource utilization on those worlds? Lunar exploration must be prepared to adjust its practices and plans if the answer is no. An alternative approach to how we interact with these environments can be found in indigenous knowledge, which is inherently interdisciplinary, multigenerational, and expressed through sustainable practices."[289]

Another paper published in 2020, this time by members of NASA's Planetary Protection Office and others ("Absolute Prioritization of Planetary Protection, Safety, and Avoiding Imperialism in All Future Science Missions: A Policy Perspective"), calls for the massive regulation of private space travel, which must, according to the authors, be placed under the primacy of "anti-imperialism."[290]

To return to the topic of colonizing Mars: In order to decide whether it is right and legitimate to try to colonize Mars, it is essential to carefully weigh up the pros and cons. The counterarguments include the "rights" of microbes and rocks, and the question of whether the religious feelings of indigenous peoples, or even the feelings of the moon itself, might be hurt.

Rubenstein and others argue that we should at least consider whether rocks do not have rights of their own. They point to the historical first moon landing in 1969, where astronauts found it difficult to ram the American flag into the ground (they hit hard rock under the dust), and interpret this as the moon trying to defend itself: "In fact, the Moon might even *desire* things. Considering the respiratory trouble it's given our astronauts and the functional trouble it's given their machines, the Moon might well be expressing a geologic desire that human beings remain on their home planet."[291]

The human colonization of Mars is not only being debated among philosophers, it has also caught the attention of NASA. Linda Billings, a consultant to NASA's Planetary Defense Coordination Office specializing in communication strategy and planning, is one of the prominent opponents in this discussion. In 2019, she wrote an article "Should Humans Colonize Mars? No," in which she argued it would be, "immoral to transport a tiny, non-representative, subset of humanity—made up of people who could afford to spend hundreds of thousands to millions of dollars on the trip—to live on Mars, as Bezos, Musk, and their advocates propose."[292]

One can only hope that the influence of anti-capitalists and proponents of woke ideology has now been permanently curtailed at NASA: After President Donald Trump signed Executive Order 14151 ("Ending Radical And Wasteful Government DEI Programs And Preferencing") on January 20, 2025, ending all DEIA initiatives in federal agencies, NASA management sent a memo to its employees two days later ordering the immediate cessation of internal DEIA programs.[293]

The above topics are currently being discussed not only at conferences and in specialist journals but also among a wider public. A debate at a conference in Reno, Nevada, later published under the title "The Great Colonization Debate," included the following statement: "Humanity itself seems to function more like an infection to the Earth than anything else, so I would ask at what point is our 'right' to survive outweighed by the rights of all the other species on Earth to survive?"[294] In the same debate, the idea of humans taking animals with them to Mars was rejected as these creatures could not be consulted or give their consent for such a journey: "If humans do go they should not bring other animals with them because the other species did not sign up for this!"[295]

Another frequently cited argument is: "It's deeply perverse to say humanity must be saved by destroying other ecosystems because it's humanity's fault we are in this situation in the first place."[296] This line of thought portrays humanity as an inherently malevolent species, responsible for centuries of misery and destruction on our planet. The general consensus of the conference was that resolute resistance to the plans to colonize Mars was needed: "We have an ethical duty to sabotage the elitist notion of colonization: As people thinking about these things, it's our job to infiltrate, convince, convert and take over."[297]

Proponents of this anti–space exploration movement praised what they see as an emerging "anticolonial spacewave." This manifested itself in the publication of a white paper in October 2020 from the Equity, Diversity, and Inclusion Working Group of the Planetary Science and Astrobiology Decadal Survey, which makes recommendations to NASA and other government agencies in the United States. These recommendations read like an anti-capitalist and anti-colonialist manifesto, with the paper stating that: "It is critical that ethics and anticolonial practices are a central consideration of planetary protection. We must actively work to prevent capitalist extraction on other worlds."[298]

The basic tenor of the paper is that capitalism, driven by its relentless pursuit of profit, has destroyed the Earth and now wants to extend this destruction to other planets. Absurdly, the claim that colonial powers conquered uninhabited lands in the past (while in reality it was often already inhabited) is now put on a par with the argument of the proponents of Mars colonization that there are no humans on the Red Planet. The paper states: "We must first reject the idea that microbial life is beyond moral consideration due to the label 'non-intelligence' or the claim that Mars is an empty place. We cannot repeat the notions of 'terra nullius' that perpetuated colonial violence on Earth."[299]

Rubenstein asserts that the rationale of those advocating for the colonization of Mars mirrors that of historical colonialists on Earth, as both groups claimed that the land they sought to colonize was uninhabited.[300] The same claim may have been made then as it is now, but it was mostly wrong then, whereas it is right now.

Yet another paper published in 2020 and authored by several individuals affiliated with NASA's Planetary Protection Office emphasizes the need to prioritize "anti-imperialism" and "anti-colonialism" in the development of legislation governing private space travel.[301]

Proponents of private space travel should not make the mistake of underestimating such efforts. Yes, some arguments, such as accusing anyone who does not recognize the "rights of rocks" of "antimineralism," do seem absurd. But history demonstrates that while entrepreneurs and engineers strive to develop new technological and economic solutions to humanity's problems, intellectuals often work to prevent them in the name of anti-capitalism. It would not be the first time that irrationality has ultimately prevailed over rationality. The consequences would be fatal if politicians influenced by such ideas ended up preventing projects such as the one spearheaded by Elon Musk.

Anyone who believes that woke ideology is dead or has lost its influence is mistaken. The history of the political left proves that they don't give up easily, and even when they appear to be defeated, they come back a few years later with even greater force. Parallel to the advances in private space travel, there has been a notable surge in opposition from anti-capitalists, advocates of woke ideology, and self-proclaimed "postcolonialists." Arguments such as advocating for the "rights" of boulders and microbes as the rightful inhabitants of Mars are obviously absurd. Especially when you consider that precisely those intellectuals for whom the protection of private property is otherwise—to put it mildly—not a priority at all, but who often even see it as the root of many evils, suddenly discover their love of property when it is not a matter of humans, but of microbes that may exist on other celestial bodies. The anthropologist Michael Oman-Reagan, speaking at the aforementioned conference on space colonization, argued: "If there are microbes in Mars, I think Mars belongs to them. If there is water on Mars, that water belongs to them."[302]

Martyn J. Fogg, a British physicist, geologist, and specialist in terraforming, rejected these and similar arguments in an article on "The Ethical Dimensions of Space Settlement" in the journal *Space Policy*: "The argument amounts to saying that humans actually have the *lowest* degree of intrinsic worth of any class of formed object. Rocks are free to rust and crumble over the aeons, asteroids and meteorites free to batter the Martian surface, and microbes free to hitch a ride if they can survive the trip and there to evolve into new forms that are Martian. Only humans should be constrained from fulfilling the evolutionary potential according to this philosophy. Yet if spacefaring is a legitimate activity for microbes, why should it not be so for humans? The allied ideologies of misanthropy and sentimentality cannot provide a satisfactory answer."[303]

Musk remains steadfast in his ambition to reach Mars—undeterred by polls and, least of all, by the criticism of left-wing ideologues. In

May 2025, he announced plans for an unmanned Starship mission to Mars at the end of 2026. "If those landings go well, then human landings may start as soon as 2029, although 2031 is more likely," Musk stated.[304] But in the meantime, he has significantly changed his schedule. Although Musk had long emphasized that SpaceX's focus was on Mars, he posted on X on February 9, 2026: "For those unaware, SpaceX has already shifted focus to building a self-growing city on the Moon, as we can potentially achieve that in less than ten years, whereas Mars would take 20+ years. The mission of SpaceX remains the same: extend consciousness and life as we know it to the stars. It is only possible to travel to Mars when the planets align every twenty-six months (six-month trip time), whereas we can launch to the Moon every ten days (two-day trip time). This means we can iterate much faster to complete a Moon city than a Mars city. That said, SpaceX will also strive to build a Mars city and begin doing so in about five to seven years, but the overriding priority is securing the future of civilization and the Moon is faster."[305]

The shift in focus has sparked speculation across the space industry. On January 13, 2026, NASA Administrator Isaacman convened a meeting with representatives from SpaceX and Blue Origin—one of his first acts as head of the agency. While details of the discussion remain scarce, it appears Isaacman encouraged the companies to share technical and organizational strategies to accelerate lunar development, emphasizing that priority would go to whichever company delivered results first.

"The effects of this meeting are already becoming apparent," notes space expert Eugen Reichl. "On January 30, 2026, Blue Origin made the surprising announcement that it would suspend activities related to the suborbital tourist spacecraft New Shepard for at least two years to concentrate all available resources on its lunar program. This is particularly remarkable given Blue Origin's recent investments in tourism flights and the fact that no fewer than four new New Shepard vehicles are currently under construction. If such a project can be paused with a single decision, it signals that the race for the Moon has truly entered a critical phase."[306]

Just days later, Musk reiterated his stance in the post quoted above, drawing a critical response from Robert Zubrin on X: "It won't work. The materials required to support a growing civilization do not exist on the Moon."[307]

I think Musk will continue to pursue Mars in the long term. For now, he appears to be aligning with NASA's immediate priorities, likely responding to concerns that China could outpace the United States in the next lunar landing.

But what about NASA's plans for Mars? There are no concrete plans, still! Harry W. Jones of the NASA Ames Research Center stated in July 2024: "It has seemed there must be overwhelming difficulties preventing our going to Mars. We have not reached Mars even though it has been NASA's horizon goal since Apollo. We do not have a detailed feasible mission plan. There has not been sufficient funding to make tangible progress. Mars plans usually propose developing advanced technology, before we can begin the mission."[308]

Between the lines, you can sense his frustration—directed not only at politicians, but also at NASA itself. For over fifty years, there have been repeated excuses for indefinitely deferring a mission to Mars. And yet, there is no shortage of plans to reach Mars. Since the 1950s, over 1,000 plans have been devised. Even if you only count those that meet scientific standards, there are still fifty-five plans.[309]

However, in a detailed analysis, Jones concludes that there are "no showstoppers"—that is, no insurmountable obstacles on the path to Mars. He identifies seven key challenges: hostile surface environment, human performance, life support, medical care, radiation exposure, reduced gravity, and telecommunication delays—and shows that there are viable solutions for each and every one of them based on the current state of technology.[310] Solutions for all these problems exist today, he argues, in part thanks to the recent great reduction in launch costs.

The primary technical challenges for a Mars mission are long-term life support, radiation exposure, and reduced gravity. According to Jones, solutions for solar flares and cosmic microwave background radiation already exist, made possible by reduced launch costs. For instance, a small, shielded shelter within the spacecraft, measuring two meters in diameter with 7.5-centimeter-thick walls, could protect against solar storms during transit. Additionally, a shorter transit time and the use of Martian regolith to shield surface habitats could mitigate the effects of cosmic microwave background radiation.

Regarding the challenges of weightlessness during space travel and the more than 60 percent reduction in gravity on Mars: A rotating spacecraft, similar to those depicted in science fiction films, can generate artificial gravity during the journey. And there are also viable solutions for habitation on Mars itself: "One suggestion is a large rotating underground wheel. The floor must be tilted so that the 0.38 g Mars gravity pulling down combines with a 0.92 g horizontal spin force to provide 1 g. The rotating wheel would be underground to provide radiation shielding."[311]

All in all, yes, a flight to Mars, and even more so its colonization, presents substantial challenges—some we can anticipate, and many others will be unforeseen. But having developed countless plans for how it could be done, the time for action has come. If Roald Amundsen and his team, the first to reach the South Pole in

1911, had required the same level of safety and perfection as NASA demands for its mission to Mars, they certainly would never have launched their expedition. Furthermore, Jones highlights another very important and very positive aspect: The recent great reduction in launch costs alone now makes it possible to solve many of the problems that previously seemed insurmountable.

CHAPTER 6

THE TREASURES OF SPACE

Have you ever come across this argument before? Because the Earth's raw materials are finite, infinite growth is impossible. This leads many to conclude that, somehow, growth must be curtailed. But how much truth is there behind this assertion?

Warnings about the limits to growth are not new; they have been around for centuries. Here are just a few examples from the last eighty years: In 1939, the U.S. Department of the Interior declared that U.S. oil reserves would last only thirteen more years. In 1949, the U.S. Secretary of the Interior announced America's oil supplies would soon run out. Having learned nothing from its earlier false claims, in 1974, the U.S. Geological Survey said that the United States had only ten years of natural gas left.

In 1970, scientists published a graph in *Scientific American* in which they estimated that humanity would run out of copper shortly after the year 2000. Lead, zinc, tin, gold and silver were expected to disappear before 1990.[312] Also in 1970, the ecologist Kenneth Watt predicted that the world would soon run out of oil: "You'll drive up to the pump and say, 'Fill 'er up, buddy,' and he'll say, 'I am very sorry, there isn't any.'"[313]

Published that same year, the Club of Rome's "Limits to Growth" study attracted a great deal of attention. To date, more than 30 million copies of the study have been sold in thirty languages. The book offered a stark message: The planet's raw materials would soon be depleted, especially oil. In twenty years, the scientists predicted, the last drop of oil would be used up. And it wasn't only in relation to oil but for almost all relevant raw materials that the Club of Rome's report completely misjudged the date by which they would be exhausted.

Natural gas, copper, lead, aluminum, tungsten: According to the predictions issued at the time, none of these natural resources would still be found in the earth today—based on forecasts for continued economic growth between the 1970s and the present day. Everything should have been used up by now; in some cases, decades ago. Silver was supposed to be depleted in 1985. In fact, in January 2020, the United States Geological Survey (USGS) estimated silver reserves worldwide at 560,000 tons.

Before anyone starts shaking their head at all these false predictions, it is worth pointing out that from the beginning of industrialization until about the 1970s, there was indeed a close correlation between economic growth on the one hand and energy and raw material consumption on the other.[314]

But based on numerous data series, the American scientist Andrew McAfee proves in his book *More from Less*, published in 2020, that economic growth has decoupled itself from the consumption of raw materials. Data for the United States show that of seventy-two commodities, only six have not yet reached their consumption maximum. Although the U.S. economy has grown strongly in recent years, consumption of many commodities is declining.[315]

As long ago as 2015, the American environmental scientist Jesse H. Ausubel confirmed in his paper "The Return of Nature: How Technology Liberates the Environment,"[316] that Americans are consuming fewer and fewer raw materials per capita. Total consumption of steel, copper, fertilizer, wood and paper, which had previously always risen in tandem with economic growth, had peaked and been declining ever since. All of these developments are due to the laws of much-maligned capitalism: Companies are constantly looking for new ways to produce more efficiently, that is, to get by with fewer raw materials. They do this, of course, not primarily to protect the environment, but to cut costs.

What's more, innovation has promoted a trend called miniaturization or dematerialization. One example of this trend is the smartphone. Just consider how many devices are contained in your smartphone and how many raw materials they previously consumed: calculator, telephone, video camera, alarm clock, voice recorder, navigation system, camera, MP3 player, compass, answering machine, scanner, measuring tape, radio, torch, calendar, encyclopedia, dictionary, foreign language dictionaries, and address book.

Most people today no longer have a fax or use paper road maps because they have everything at their fingertips on their smartphones, and some even do without a wristwatch, using the clock on their iPhone instead. In the past, you had four separate microphones in your telephone, audio cassette recorder, Dictaphone and video camera. Today, you use just a single microphone in your smartphone.

When I was young, I used to be proud of my large record collection, which spanned several shelves. As technology advanced, I bought CDs that all fit on a single shelf—and consumed far fewer raw materials. Today, my girlfriend teases me because I still buy CDs—all of her music is in digital files, which don't take up any space at all. I admit I'm a bit old-fashioned and own several thousand books, but many now prefer to read e-books or listen to audiobooks on their smartphones. These are just a few of the many examples that demonstrate the trend toward dematerialization, which contributes to a lower consumption of raw materials.[317]

In 2025, the physicist and energy economist Björn Peters wrote about "finite resources"[318] and identified three hierarchical levels of resources—natural, technological, and economic:

Natural resources are all found in nature—wind, sun, water, rocks, and plants. These become technical resources only when humans possess the technologies to exploit them. For instance, oil became usable with the invention of the kerosene lamp and the internal combustion engine, and sunlight became harnessable through the photoelectric effect. However, whether a technically accessible resource is actually utilized depends on the cost of its extraction or exploitation relative to its benefit—thus transforming it into an economic resource.

The economic viability of resources is influenced by fluctuations in prices, demand, and technological advancements. Falling extraction costs or rising prices make new deposits profitable. In the case of crude oil, for example, recoverable quantities have increased thanks to processes like enhanced oil recovery, which could now make up to 30 percent of original oil reserves recoverable. With technologies such as in situ cracking, it might even be possible to utilize 75 percent of these reserves.

Statements about the remaining lifespan of raw materials—often cited as twenty-five to thirty years—are not real limits, but rather economic ones. Companies only report reserves that can be extracted economically with current technology and at prevailing prices. These are referred to as "measured" or "determinable" resources, while "presumed" deposits are not included in companies' accounts. And, since companies typically only plan about twenty to thirty years ahead, reserves always appear limited; in reality, they are continually being expanded through innovation. In particular, according to Peters, exploration costs for new raw materials are steadily decreasing due to technological advancements, making it economically unjustifiable to prospect for raw materials more than twenty to thirty years ahead.

Raw materials, therefore, do not run out; they merely become more or less expensive or more important or are replaced. Human creativity plays a pivotal role in this dynamic. Sustainability cannot mean stagnation, but rather faith in progress. Historical examples such as salt or yew wood show that scarcity always passed when new technologies emerged.

The reality is therefore more complex than it might initially appear when people say, "Because the Earth's raw materials are finite, infinite growth is impossible." Humanity consumes only a minuscule fraction of Earth's mineral resources, such as copper, iron, aluminum, and phosphorus. For copper, the most extensively used resources, annual consumption is approximately 0.000034 percent of the available supply.[319]

Moreover, the argument that limited resources constrain global economic growth is flawed for another reason. Why should we only consider Earth-bound resources? Beyond Earth, there are abundant resources on the Moon and, more significantly, on asteroids. The asteroid belt between Mars and Jupiter alone is estimated to contain between 700,000 and 1,700,000 asteroids with diameters of at least 0.6 mile.

In addition, there are over 30,000 known near-Earth asteroids (NEAs), and about 2,000 to 3,000 more are discovered each year.

Although they are called near-Earth asteroids, very few come close enough to Earth to pose a threat. For example, some asteroids in the Amor group are, at their closest, 28 million miles from Earth. Most NEAs have diameters of less than one mile. The largest near-Earth asteroid, Ganymede (not to be confused with Jupiter's moon of the same name), measures twenty miles in diameter and can come as close as 32 million miles from Earth.[320]

When I was invited by the Federation of German Industries in 2024 and 2025 to speak at a raw materials and space congress on the topic of "Asteroid Mining," my audiences' reactions were divided: Many were curious, but many were also skeptical. This was, in part, due to the numerous misunderstandings about asteroid mining.

Asteroid mining has always been a polarizing topic. Nearly four decades have passed since Jeff Bezos addressed a group of students at Princeton University and speculated on the quickest way to make asteroids usable for humans. During the discussion, one student jumped up, shouting angrily: "How dare you rape the universe!" before storming out of the room. All eyes turned to Bezos, who asked: "Did I hear her right? Did she really just defend the inalienable rights of barren rocks?"[321] The American religious scholar Mary-Jane Rubenstein, whom we met in the last chapter, professes her sympathy with the student and empathizes with her anger. The same critics who condemn capitalism for leading to the depletion of Earth's resources vehemently oppose the use of resources from other celestial bodies.

Aside from these ideologically based objections, the question arises: Is asteroid mining even economically viable? Isn't it far too expensive to transport raw materials from asteroids to Earth?

The media sometimes covers stories about asteroids, such as Psyche, an asteroid with a diameter of about 140 miles. Psyche is between 185 and 375 million miles from Earth. NASA launched a probe—also called *Psyche*—which is en route to the asteroid and is scheduled to arrive there in August 2029. Once it arrives, the probe will study the asteroid for twenty-six months in four different orbits to analyze its geology, topography, elemental composition, magnetic field, and mass distribution.[322]

The figures quoted in the media about the possible value of asteroids like Psyche are so high that I find myself adding how many zeros they have in brackets. In one of the many newspaper articles, headlined "NASA plans mission to a metal-rich asteroid worth quadrillions," scientists speculated about a rapid collapse of the world economy if the wealth of Psyche was tapped: "It's such a strange object," said Lindy Elkins-Tanton, the lead scientist on the NASA mission and the director of Arizona State University's School of Earth and Space Exploration. "If the two-hundred-kilometer-wide body could somehow be transported back to our planet, Elkins-Tanton has calculated that the iron alone would be worth US $10,000 quadrillion [= 10,000,000,000,000,000,000 → 19 zeros]. The gross world product (GWP) in 2015 was only about $73.7 trillion [= 73,700,000,000,000 → seventeen zeros], so our economy would promptly collapse."[323]

There were also sensational reports about the asteroid Germania, with media reports valuing it at US$100 trillion, which at the time was more than the entire gross domestic product of the planet Earth.[324] There is even a dedicated website that lists the purported values of asteroids—and a search for "most valuable" reveals countless entries with values exceeding US$100 trillion.[325]

Martin Elvis, a renowned astrophysicist at the Center for Astrophysics, Harvard & Smithsonian in Cambridge, who has published over four hundred papers primarily on supermassive black holes (quasars), ridicules such claims: "The paeans to the value of space resources have mostly been written by scientists, not business people. And we scientists are pretty financially naive."[326]

If valuable metals such as platinum flooded the market in such quantities, the price would of course immediately fall dramatically. In addition, the composition of Psyche and other asteroids has so far only been assessed through remote sensing and spectral analysis. Recent research has concluded that Psyche, for example, may not be as rich in metals as previously thought.[327] Precise data will only be available after an on-site investigation or by taking samples.

Nonetheless, it is widely acknowledged that certain asteroids contain valuable raw materials, such as platinum group metals (PGMs), in concentrations significantly higher than those found on Earth. Of the ten most valuable raw materials in the world, around five to six belong to the platinum group, and the others (e.g., gold) are probably also found in much larger quantities and are more easily accessible on asteroids than on Earth.

The metals on asteroids are roughly divided into two groups: precious metals, such as the platinum group metals, which could be transported to Earth for sale, and base metals, such as aluminum (Al) and silicon (Si), which could be utilized directly in space for constructing space stations or solar modules.

In a paper published in 2023 on "Precious and structural meals on asteroids," scientists reported that there are only a few elements that occur in asteroids in higher concentrations than in the best ore deposits on Earth. However, these include the very valuable platinum group metals, which are found in iron meteorites in quantities between around 6 and 230 ppm (= parts per million)—and therefore more than in almost all ores on Earth. Earlier estimates of up to 300, or even 700 ppm, however, have not been confirmed according to the latest measurements.[328] The same researchers also emphasized that so-called M-type asteroids were previously thought to be large chunks of metal. Today, however, it is known that they often consist of a mixture of silicates and metals.[329]

The fact that the concentration of PGMs is many times higher, at 6 to 230 ppm, than in the Earth's crust is because most asteroids never experienced the process of 'differentiation' that took place on planets such as the Earth.

The Earth lost most of its siderophile elements to the core during planetary differentiation—a process in which an initially homogeneous mixture of molten rock and metal separated into layers under the influence of gravity, causing dense, iron- and metal-rich material to sink toward the center while lighter silicate rocks rose toward the surface. Some large bodies in the early solar system underwent similar differentiation but were later shattered by massive collisions. The fragments of their exposed metallic cores survive today as certain metal-rich asteroids, containing high concentrations of iron, nickel, and platinum-group elements that are rare in Earth's crust.

"This means," explains the leading expert on asteroid mining, John S. Lewis, "that the terrestrial concepts of having to remove vast masses of uninteresting overburden, and of ore veins that must be found and followed deep within hard rock, are largely irrelevant to asteroid mining. Undifferentiated asteroids consist of a fine-grained mixture of all the materials that went to make Earth's core, mantle, and crust. . . . As an example the concentration of the 'rare' and 'precious' platinum-group metals in a random piece of an average asteroid is several times higher than in the richest known ore bodies in Earth's crust!"[330]

Other common misconceptions about asteroids also persist. Many think that asteroids are too far away, that the distances are too great.

This is true for asteroids in the asteroid belt between Mars and Jupiter, at least with today's technological capabilities. The Psyche space probe, which I mentioned above, was launched on October 13, 2023, and is expected to reach its target asteroid in July or August 2029. This means that the journey will take around six years. A mission to the asteroid and the subsequent transportation of the platinum to Earth would take more than a decade. However, a mission launched from Mars would be much shorter and would only have to overcome the 60 percent lower gravity of the red planet during takeoff.

Lewis estimates that there are around 3,800 near-Earth asteroids (NEAs) that require less fuel to reach than the Moon. Although the Moon is relatively close to Earth, at a distance of around 240,000 miles, the energy required to get there is comparatively high. In contrast, some NEAs follow orbits that closely resemble Earth's, making them more accessible for exploration—even though the asteroids are often farther away than the Moon. This apparent contradiction is explained by the fact that it is not only the distance that counts, but above all the relative energy expenditure. However, there is another problem that Lewis highlights: "The near-Earth objects present a very different set of problems: The richness of NEO resources cannot rationally be doubted. Typical desirable resources are hundreds to thousands of times more abundant in NEIs than on the Moon. But the round-trip times for missions to NEOs are typically two to five years—not one week, as with the Moon."[331]

At first glance, this seems paradoxical, but it can be explained as follows: Although, from an energy-calculation standpoint, some near-Earth asteroids require less energy to reach than the Moon, missions to them generally take significantly longer. The Moon orbits the Earth at an approximately constant and near distance, making it easily accessible at almost any given time; travel to the

Moon takes only a few days. Asteroids, by contrast, follow elliptical orbits around the sun that are more or less inclined to the ecliptic. As a result, they are not always close enough to be accessed from Earth and are only reachable within a reasonable time and energy budget during certain favorable orbital alignments. Any return flight to Earth would also be similarly contingent on a favorable launch window. While you can fly directly to the Moon and return almost at will, if you flew to an asteroid, you would often have to wait many months for a good opportunity to return. In addition, the trajectories that minimize energy consumption are inherently longer because they follow curved rather than direct paths. All of this means that missions to near-Earth asteroids will invariably take much longer than missions to the Moon.

Another common objection to asteroid mining is that it would cost too much to transport raw materials from an asteroid to Earth, rendering mining economically unviable. The fact that transporting materials is expensive is partly true, but the argument still misses the point. Experts such as Lewis, who raised the issue as early as 1997 in his book *Mining the Sky*, argued above all in favor of using the raw materials extracted from asteroids in space: "The resources in greatest demand for use in space are those that can be used to meet a space civilization's most massive material and energy needs. The cost of transportation from Earth is so large that any material needed in large masses in space, almost irrespective of its market price here on Earth, should—whenever possible—be made from resources found in space. Another corollary of this principle is that only materials with a high unit market value on Earth (thousands of dollars per kilogram) may be worth transporting back to Earth."[332]

Lewis wrote this in 1997, when the cost of transporting payloads into space was much higher, but in principle it still applies. For example, in many cases it will make more sense to use raw materials from asteroids in space to build space stations or large solar panels than to transport them there from Earth.

In particular, water can be extracted from many asteroids. Astrophysicist Martin Elvis even believes that water extraction could possibly become more important than the extraction of raw materials such as platinum. Transporting water from Earth into space remains prohibitively expensive, even with reduced launch costs, yet it is an essential resource for activities such as rocket fuel production. Water can be split into oxygen and hydrogen and then used as fuel.

In future, spaceships will have to be refueled in space. Just as motorized vehicles on Earth rely on filling stations, space travel will depend on refueling stations. Running a rocket fuel station could prove to be a lucrative business opportunity. This fuel could be used for a number of purposes, including repositioning large and expensive communications satellites or clearing hazardous space debris. Water can also be used as an excellent shield against cosmic radiation in space. Research indicates that water offers similar or better protection than many other materials of comparable mass. Moreover, water is an essential resource on spacecraft anyway (e.g., as drinking water, a coolant, or as a component of life-support systems). Water could be stored in walls and tanks, simultaneously providing radiation protection.[333]

This is the crucial point: I have found in many discussions with people who had never really explored the subject (but nevertheless held firm opinions on it) that one basic assumption was fundamentally wrong, namely that the primary purpose of extracting raw materials from asteroids is to bring them back to Earth. While this may be the case for extremely valuable materials like rare earths, the majority of raw materials will actually be utilized for production in space. Even with further potential reductions in launch costs from Earth, it will often be more cost-effective to utilize raw materials obtained from asteroids in situ. In years to come, space-based production will play a significant role. Why should we transport raw materials from Earth into space when ample resources are available on planets, moons, and asteroids?

Another misconception is perpetuated by films such as *Armageddon*, in which Bruce Willis and his team transport super-heavy drilling machines onto an asteroid. Popular science fiction has distorted our ideas about the nature of these celestial bodies. However, on asteroids, as John S. Lewis explains, digging tunnels or stripping off overburden will rarely if ever be relevant. Because of the wide compositional variability of, and enormous strength differences between, different meteorite groups and their corresponding asteroid families, the appropriate mining techniques are likewise different.[334]

Many asteroids are even referred to as flying "rubble piles," as the material hardly sticks together due to the almost complete lack of gravity. In a paper called "Rubble-Pile Near Earth Objects: Insights from Granular Physics," expert Karen E. Daniels emphasizes that most near-Earth asteroids consist of "fractured rock, sometimes highly fractured and porous, and they have come to be known as rubble piles . . . weakly held together as an aggregate by a combination of both gravitational and van der Waals forces."[335] Van der Waals forces are weak physical forces of attraction between molecules or particles that are not the result of chemical bonding.

In many cases, crushing rocks with heavy equipment is not even necessary. On the contrary, the real problems lie elsewhere, such as dealing with the challenges of landing on small piles of rubble, or finding ways to securely anchor a spaceship or mining equipment. Will the asteroid remain stable, or could it break up and become insecure during activities such as landing, anchoring, or mining?[336]

As Lewis points out, docking with an asteroid can be very difficult, especially with smaller asteroids: "Asteroids, especially those smaller than several hundred kilometers in diameter, are often extremely irregular in shape. Their surfaces are a chaotic patchwork of craters, rocks, rubble, and dust. Some areas may be large expanses of stainless steel; others may be deep loosely packed regolith containing rocks of all sizes, and some may be extremely fragile and unstable 'fairy castles' of fine-grained dust. In addition, asteroids rotate on their own schedule, not in accord with pre-planned mission guidelines. The overwhelming majority of NEAs, those that are less than 100 m in diameter, are collision fragments that were originally produced in violently disruptive impact events. They often have rotation periods of a few minutes. Many are so irregular in shape that they partake of tumbling motions that cannot be described in terms of rotation around a single well-defined axis."[337]

These tasks will be more practicable on larger asteroids. And when mining starts, there's also the question of how to stop materials from floating off into space due to the almost complete lack of gravity. However, solutions to this problem do exist, such as the cable-cutter-bag mining concept, which was developed by NASA in the 1990s to address the difficulty of taking samples in an extremely low-gravity environment without the need for complex, heavy, or risky technologies such as drills or anchor systems.

The cable-cutter-bag method involves the use of a net or bag that is placed or stretched over part of the asteroid's surface and held in position with the help of cables. Once the target material—such as loose rock—is enclosed by the net, special cutting devices are used to cut specific cable connections, causing the net to contract automatically and enclose the trapped material.

What makes this method so special is that it does not require any physical anchoring to the ground. And this is particularly important in the microgravity of an asteroid, where fixing a spacecraft firmly to the surface is difficult due to the porous nature of the surface. Even the smallest forces, such as those generated during drilling, could repel or destabilize the entire system. The cable-cutter-bag concept effectively avoids this problem. In addition, the system is technically simple: Cables, nets and cutting mechanisms are lighter, more robust and less prone to failure than many mechanical alternatives. The risk of material loss is also low, as the rock remains in the sealed net after collection and cannot escape into space. I was amazed to find there is a very detailed explanation of this and other asteroid mining techniques in a 2019 book that deals exclusively with this topic, *Space Mining and Manufacturing. Off-World Resources and Revolutionary Engineering Techniques*.[338]

Some asteroids could even be used after mining is complete, as experts suggest: "After finishing the mining process and the laser sintering [process of fusing material with laser, R.Z.] of the inner surface the remaining hollow asteroid can be used as a shelter for industrial facilities or for the storage of products like water, oxygen or other gases. The rock hull provides shelter against micrometeorites, cosmic ray, solar flares and last but not least thermal insulation. NEAs with more than 400 m diameter can be used for *human colonies* with artificial gravity."[339] Their rotation creates an artificial gravity.

In order to mine asteroids more effectively, there have also been discussions about transferring asteroids into orbit around the Moon. However, there are risks associated with such endeavors. One significant concern is the possibility of altering an asteroid's orbit during the mining process, which in the worst case would lead to a collision with the Moon or even the Earth. This risk, although small, should not be taken, as it could provide fodder for those opposed to asteroid mining for completely different reasons to launch scaremongering campaigns.[340]

In an article discussing the economic aspects of asteroid mining, Mike H. Ryan and Ida Kutschera write: "Many practical problems facing asteroid mining have clear and unambiguous solutions. . . . The question is not whether asteroids can be mined, but whether or not the financial return warrants the extensive investment necessary for such mining operations."[341] They highlight the high-risk nature of asteroid mining as an investment, and caution that investors must also be prepared to accept long time horizons until they see any return on their investment.[342]

Several companies in this segment have already failed, such as Planetary Resources and Deep Space Industries. Planetary Resources, founded in the United States in 2012, initially garnered significant attention and secured notable investors such as Google cofounder Larry Page, longtime Google CEO Eric Schmidt, and movie director James Cameron. The company planned to use small satellites to identify asteroids and later mine water and platinum. Despite its high-profile backers, Planetary Resources ultimately failed to secure the funding it needed: The planned investments did not materialize, the technical challenges were greater than expected, and there was no robust business model with a medium-term earnings outlook. In 2018, the company was acquired by the blockchain company Consensys, which chose not to pursue Planetary Resources' original business model.

Similarly, in 2013 Deep Space Industries (DSI) was also founded in the United States and pursued comparable objectives, in particular the extraction of water from asteroids for use as space fuel. In contrast to Planetary Resources, DSI was an early mover in the development of small satellite platforms and planned to generate revenue from these technologies. Nevertheless, the hoped-for technological breakthroughs failed to materialize, and the company encountered similar challenges to Planetary Resources, including technological barriers, uncertainty regarding economically feasible implementation, and a lack of investor support. As a result, in 2019, the company was acquired by Bradford Space, leading to the discontinuation of its asteroid mining business. Although DSI had a lower media profile, it was considered, alongside Planetary Resources, as one of the two key companies that wanted to advance the concept of an extraterrestrial commodities market in the private space exploration sector.

So, what happens next? First of all, it will be important to collect samples from as many asteroids as possible in order to gain a better understanding of their composition and to pinpoint those celestial bodies—preferably NEAs—that are the most promising candidates for future mining operations.

So far, our understanding of asteroid composition is largely based on the analysis of meteorites. In the early 1970s, a researcher discovered that the spectral signature of Vesta, the second-largest asteroid in the asteroid belt, was almost identical in composition to that of a meteorite. This finding led to the hypothesis that the composition of meteorites could also offer insights into the composition of asteroids. Therefore, if the spectral analysis of an asteroid aligns with that of a particular meteorite, it can be inferred that the asteroid shares a similar composition.

"That is why meteorites are often defined as 'poor man's space probes'—with tens of thousands of samples collected, analyzed, and catalogued, we have gained a good understanding of the composition and geology of asteroids even though only a handful of them have been visited and studied by robotic probes. In the absence of exploring in situ or returning samples to Earth, the only way to determine the composition of a given asteroid is by analyzing the sunlight reflected from its surface."[343] However, the majority of meteorites that hit Earth originate from NEAs, so the knowledge gained from their analysis may not always be directly applicable to asteroids within the asteroid belt.[344]

The further exploration of asteroids is worthwhile in several respects. In the case of near-Earth asteroids in particular, it is important to note that they not only pose a potential threat to our planet if they are on a collision course but also offer economic potential due to their mineral composition. Therefore, asteroid missions serve a dual purpose: They both facilitate the development of strategies to defend against potential impacts and the investigation of valuable resources. The importance of such research cannot be overstated, as it promises both security-related and long-term economic benefits.[345]

To date, only a handful of missions have successfully collected samples from asteroids and brought them back to Earth. Here are the most notable missions:

Hayabusa (Japan, 2003–2010):
Asteroid: Itokawa
Mission: The Japanese space probe *Hayabusa* was the first to bring samples from an asteroid to Earth. It landed on the asteroid Itokawa in 2005 and took small samples. Despite encountering technical challenges, the probe returned with approximately 1,500 particles of asteroid material weighing 1 microgram.
Hayabusa2 (Japan, 2014–2020):

Asteroid: Ryugu
Mission: The successor mission, *Hayabusa2*, also carried out by the Japanese space agency JAXA, landed on the asteroid Ryugu and collected samples, including material from the asteroid's subsurface. In 2020, *Hayabusa2* returned to Earth with 5.4 grams of asteroid material.
OSIRIS-REx (NASA, 2016–2023):
Asteroid: Bennu
Mission: NASA's OSIRIS-REx mission, launched in 2016, reached the asteroid Bennu in 2018 and performed a "touch-and-go" (TAG) maneuver in October 2020 to collect samples. OSIRIS-REx returned to Earth in September 2023 and brought back approximately 250 grams of material—the largest extraterrestrial sample brought back to Earth from an asteroid to date.

We can already look forward to further such explorations that will give us more certainty about how much of which elements can be found in which asteroids.

The progress in rocket technology described in the previous chapters has had a major impact on the question of profitability. In 2013, the astrophysicist Martin Elvis, quoted above, estimated the number of near-Earth asteroids that could be economically viable to exploit at just ten—based on an equation in which basically all variables were unknown and the result was only a very vague estimate.[346]

In part, the development of more powerful rockets by SpaceX has greatly enhanced the accessibility of potentially valuable asteroids. Consequently, Elvis has since substantially revised his initial estimate. In his 2021 book, he states: "Putting all the factors together, we could well go from just ten precious-metal ore-bearing asteroids in our catalogs in 2013, when I first estimated the number, up to a couple of thousand within a decade, and maybe many more for water. Each one of them would be worth $1 billion or more. That's beginning to sound like real money."[347]

The primary task of astronomers and researchers deploying probes to these asteroids is now to identify the right asteroids—everything depends on it. "If we choose poorly, we'll be out of business,"[348] says Elvis. Once promising asteroids have been identified, a race to exploit them could begin. And this raises another question: Is there even a legal basis for such activities? If not, nobody will do it. The Outer Space Treaty of 1967, ratified by 116 nations, prohibits the ownership of "celestial bodies" (although this term is not defined and some legal experts have doubts as to whether it applies to smaller asteroids). But the U.S. Commercial Launch Competitiveness Act of 2015 allows Americans to engage in the commercial exploration and exploitation of space resources. More on this in chapter 10.

It is pleasing to note that public reservations about asteroid mining appear to be less pronounced than about mining on the Moon, in the deep sea, or in Antarctica. This is shown by a survey of 6,239 respondents in twenty-seven countries on five continents published in 2022.[349] But it doesn't have to stay that way. If asteroid mining goes from being a fringe issue that only a few experts are concerned with to a practically relevant topic of public debate, then anti-capitalists will also take up arms against it and influence public opinion accordingly.

The American anti-capitalist Brian Merchant has already written about this: "Rich investors from rich companies living in rich countries will benefit first and foremost—asteroids may add 'tens of billions' of dollars to the global GDP as Planetary Resources claims, but guess where most of that is going? Hint: the same place where profits from the sale of commodities have gone since the dawn of human civilization—the pockets of the 1 percent. . . . Not content with the finite supply of materials on our planet, we're searching out new celestial bodies to pillage. We are expanding the capitalist horizon into space, we are pushing the profit-motive beyond orbit."[350]

Anti-capitalists, who always talk about "sustainability" and protecting the environment, oppose asteroid mining, which could be an important alternative from an environmental perspective. In a paper published in 2023 called "Mining in Space Could Spur Sustainable Growth,"[351] scientists show that mining metals in space could be a way to make economic growth more sustainable. If it were possible to mine metals on asteroids or the Moon, that could mitigate several challenges associated with terrestrial raw material extraction. On Earth, the discovery of new, economically viable, high-quality ore deposits is becoming increasingly rare, meaning that ever-larger quantities of rock need to be processed to extract the same amount of metal.

Space, in contrast, offers access to metal-rich bodies that could be exploited with fewer ecological consequences, so the argument goes. As advancements in space travel continue to reduce transportation costs, space mining could become economically attractive. At the same time, this would decrease reliance on environmentally harmful processes on Earth, significantly reducing CO_2 emissions associated with energy-intensive metal extraction.[352] And finally, as Elvis also points out, space exploration may lead to the discovery of entirely new materials or raw materials that do not exist on Earth, but for which innovative applications will be found.[353]

CHAPTER 7

SPACE AS A TOURIST DESTINATION

One problem with private space travel is that it has thus far been overly reliant on subsidies or contracts from governments. We can distinguish between three distinct phases:

- **Phase 1:** Government-run space exploration (1957 to around 2010): During this period, space exploration was predominantly managed by government space agencies, with minimal private sector involvement.
- **Phase 2:** Private and public sector space travel (from around 2010 to the present): This phase marks the growth of private spaceflight, although subsidies and government contracts, such as those from NASA and the military, continue to play an important role, alongside private investments, which have become much more important. A significant source of additional financing has been the substantial funds provided by billionaires such as Jeff Bezos, Elon Musk, Paul Allen, and

Richard Branson. However, these funds were largely (in the cases of Jeff Bezos, Paul Allen, and Richard Branson) or significantly (in the case of Elon Musk) earned in other economic sectors outside the space industry.

- **Phase 3:** Predominantly private space travel (future): In this phase, private spaceflight will be primarily financed by the revenues it generates.

In the previous chapter, I explained how space mining could be a future source of income, and I will deal with other revenue streams, particularly from the satellite industry, in the next chapter. In chapter 10, I will outline an extremely important source of future revenue, namely from trading real estate—although this is contingent upon the recognition of private property as a foundation for space development. Another potential source of revenue that has not yet been mentioned could be space tourism, which began a quarter of a century ago, in 2001, with flights to the ISS.

The field of space tourism has generated an astonishing volume of scientific literature. In 2025 alone, a survey of the scholarly literature on space tourism cataloged 299 scientific publications, books, and other resources—more than the number of actual space tourists to date.[354] The list of articles extends to topics such as "Designing a Zero Gravity Toilet for Disabled Space Tourists." It seems that no aspect of space tourism has been ignored.

In the past, a number of people flew into space even though they were not professional astronauts. For instance, Christa McAuliffe was a participant in NASA's "Teacher in Space Program." The program aimed to promote public interest in space exploration and McAuliffe, a civilian teacher, was going to conduct lessons from space to inspire students. On January 28, 1986, she and six other crew members boarded the space shuttle *Challenger*. Just seventy-three seconds after takeoff, the shuttle exploded due to a seal failure on a solid fuel booster caused by the low temperatures on launch day. All seven astronauts, including Christa McAuliffe, perished in the accident.

Then there was the Japanese journalist Toyohiro Akiyama, who embarked on a space mission four years later aboard the Russian Soyuz TM-11 on December 2, 1990, alongside two Russian cosmonauts. Like McAuliffe, Akiyama did not pay for his flight himself, he was on a professional assignment. The Russians officially referred to him as a "research cosmonaut," but he was actually on a marketing mission. His employer, the television station TBS, funded his place on the mission—at a reported cost of between $10 and $14 million (equivalent to approximately $23 to $32 million in today's money).[355] During his eight-day mission, six of which were spent on the Soviet space station *Mir*, he was tasked with providing daily reports on his impressions and experiences in space to commemorate the 40th anniversary of TBS. But he suffered severely from space sickness—a condition similar to seasickness caused by a disturbance in the sense of balance—and nicotine withdrawal symptoms because he was a chain-smoker who normally smoked four packs a day.[356] His reports were boring ("I'm looking out of the window right now") or bizarre ("Fat Japanese frogs in space love weightlessness. Thin Japanese frogs behave as if they would rather be in Yokohama"). The TV station's ratings were not much higher after the PR mission than before.[357]

Dennis Tito, who journeyed to the ISS in April 2001, is widely recognized as the first true space tourist to pay for his flight out of his own pocket. The American entrepreneur and former NASA engineer traveled aboard a Russian Soyuz rocket. The flight was organized by the Russian space agency Roskosmos in collaboration with the private U.S. company Space Adventures. Tito spent a total of around eight days in space, including nearly seven days on the ISS. He paid $20 million for the experience (equivalent to about $34 million today).

Initially, it wasn't entirely clear whether his trip would be allowed to go ahead: NASA cited safety concerns because Tito was not a professional astronaut. However, Russia prevailed and went ahead with the mission. During his time in space, Tito conducted small scientific experiments and extensively filmed for personal purposes. After his return, he expressed his enthusiasm for the experience and predicted that space tourism could evolve into a mass market in the future.

Between 2001 and 2009, seven individuals traveled to the ISS as paying space tourists on Russian Soyuz spacecraft via the Space Adventures agency. They underwent months of rigorous training, and even had to learn Russian—the reason given was that they needed to understand the language in order to be able to implement emergency measures if necessary.[358]

After 2009, there was a hiatus and private visits to the ISS were no longer possible. Since 2022, however, private individuals have resumed visiting the ISS, facilitated by Axiom Space using SpaceX spacecraft. Tourists now pay approximately $55 million per person for such trips.

Interestingly, the much more complex and expensive orbital flights were the first undertaken by space tourists, while the comparatively much less expensive suborbital flights, which do not escape Earth's gravitational field, only came later. There are also major differences between the various suborbital flights on offer, as we will see from the examples of Blue Origin (Jeff Bezos) and Virgin Galactic (Richard Branson):

In the case of Blue Origin, the New Shepard rocket transports a capsule beyond the sixty-two-mile (100 km) mark, the so-called Kármán line, where it separates from the capsule, which then returns to Earth via parachute. During this fully autonomous, pilotless flight, passengers experience a few minutes of weightlessness. In contrast, Virgin Galactic's VSS *Unity* launches horizontally from a carrier aircraft, White Knight Two, which releases the spacecraft at a high altitude. VSS *Unity* then ignites its rocket engine, ascending to between fifty-three and fifty-six miles, which is below the Kármán line but above the U.S. space boundary of fifty miles (we will come back to this definition shortly).

In 2021, significant media attention focused on the space race between the two billionaires Jeff Bezos and Richard Branson. Branson, founder of Virgin Galactic, unexpectedly announced that he would be flying into space onboard the VSS *Unity* on July 11, 2021—nine days ahead of the Amazon boss Jeff Bezos, who had previously scheduled his Blue Origin New Shepard flight for July 20, 2021. Branson traveled with three other passengers and two pilots on board his spaceship, which was brought to an altitude of approximately ten miles by a carrier aircraft before igniting the engines of its rocket module to climb to an altitude of around fifty-54 miles. Bezos flew with his brother Mark, an eighty-two-year-old former astronaut candidate, and an eighteen-year-old passenger, reaching an altitude of sixty-six miles. Although both flights lasted only a few minutes, they held significant symbolic value in the billionaires' race for commercial space travel. Branson became the first entrepreneur to fly into space with his own space company, at least if you accept the U.S. definition that space begins at an altitude of fifty miles. Bezos was the first if you use the international definition of sixty miles.

Branson got into the airline business during a vacation when his return flight was canceled. Without further ado, he chartered a plane for $2,000 and divided that by the number of passengers. He borrowed a blackboard and wrote on it: "VIRGIN AIRWAYS. $39 SINGLE FLIGHT TO PUERTO RICO."[359] It worked so well that he later had the idea of founding his own airline. Then, in 2004, he founded Virgin Galactic with the ambitious goal of making space travel more accessible. However, the company faced significant setbacks, including a fatal accident on the ground in 2007 in which three employees lost their lives. In 2014, SpaceShipTwo VSS *Enterprise* crashed during a test flight, killing the co-pilot and seriously injuring the pilot.

These tragedies led to many years of delays, soaring development costs, and a loss of confidence among investors. Despite an IPO in 2019 and a successful flight in 2021, Virgin Galactic remained loss-making and struggled with a series of launch delays and financial problems. In spring 2024, the company made the decision to discontinue its commercial flights with the spaceplane *Unity* and shift its focus to the further development of the launch system.

In September 2021, two months after the duel of the space billionaires, the crew of a mission called Inspiration4 flew much higher than Branson and Bezos had managed. This trip, which was also financed by a billionaire, lasted three days and took four private astronauts into orbit at an altitude of around 365 miles—higher than the ISS. Inspiration4 marked a significant departure from previous spaceflights as it did not have any professional astronauts on board. The commander, Jared Isaacman, is the billionaire founder of Shift4 Payments and was subsequently appointed administrator of NASA on December 18, 2025. He owns what many describe as the largest fleet of fighter jets outside the regular air force.

Inspiration4's crew included Hayley Arceneaux, a bone cancer survivor, and two other private individuals who were selected through a fundraising campaign for St. Jude Children's Research Hospital that raised $240 million—with more than half of the donations coming from Isaacman and Elon Musk. Technically, it was a fully autonomous Dragon capsule mission controlled by SpaceX, without the need for docking to a space station.

Space tourism is slowly taking off. Blue Origin does not publish its prices, but they are said to range from $200,000 to over $1 million.[360] And space tourism will not stop at suborbital flights and trips to space stations. In 2018, Elon Musk revealed that the Japanese billionaire Yusaku Maezawa would be the first space tourist to orbit the moon with SpaceX. The voyage, planned for 2023, was to last a week aboard the Big Falcon Rocket (later Starship). Maezawa, the founder of the online fashion retailer Zozotown, secured his place on the rocket by paying a large, undisclosed sum for the "dearMoon" mission. He planned to bring along six to eight artists at no cost, including DJ Steve Aoki, to inspire their creativity. In 2020, Maezawa sought a female companion for the trip through a public tender and in 2021, he opened applications for potential participants worldwide. However, due to delays in the development of Starship, the proposed schedule proved to be unachievable. Consequently, in 2024, Maezawa canceled the Moon trip as the launch date remained uncertain. SpaceX continued to prioritize rocket development and other missions, such as resupply flights to the ISS.

Looking ahead, it is obvious that flights around the Moon will represent the next frontier for space tourism. As settlements on the Moon become a reality in the coming decades, a burgeoning tourism industry will take shape. As Christopher Wanjek writes in his book *Spacefarers*: "Tourism is certainly in the Moon's future. One can easily envision a two-week trip to the Moon to be much like an African safari was 150 years ago: initially for the wealthy, with a tinge of danger, and certainly not for the kids, at least not at first."[361]

Hotels will be built on the moon, and the two big attractions will be weightlessness and the unobstructed views into space: "There's certainly a wow factor in standing on the edge of a massive lunar crater and thinking about the impact that created it. But, the real beauty on the Moon may be the Earth, which would appear as a massive orb in the sky, up to six times larger than our view of the Sun or Moon from Earth. From a lunar perspective, the Earth will wax and wane through the course of a month, and swirling cloud patterns and large storms would be visible from your hotel observation deck."[362]

The lunar environment, with gravity six times lower than that of Earth, will offer unique conditions for sporting events (e.g., athletics)—and plenty of opportunities for money-spinning TV broadcasts to Earth and lucrative sponsorship deals. But this is all probably a few decades in the future. Are we witnessing the birth of a new industry, space tourism? Or will it just remain an expensive hobby for the very rich? Is the situation perhaps similar to the early days of aviation?

The first aviation companies emerged in the 1920s, followed by the development of commercial aviation. At the time, however, flying was only accessible to the wealthy. Even in the early 1960s, a first-class transatlantic flight cost as much as a new small car. Cheap flights as we know them today, at the kinds of prices everyone can afford, simply did not exist back then.

It is not unusual, in fact it is the rule, that only the rich can afford new products at first. It is these affluent consumers who finance the development costs of products before they are ready for mass production. In the Middle Ages, glass windows were a luxury that only churches and palaces could afford. It was not until the nineteenth and twentieth centuries that glass windows became affordable for the general public. Flush toilets were developed in England in the eighteenth century, but remained a privilege of the upper classes for a long time. In many European households, bathrooms with toilets only became the norm after World War II. Upon its invention in the late nineteenth century, the automobile was an expensive, handcrafted item that only the rich could afford. It wasn't until Henry Ford's assembly-line production began in 1913 that the cost dropped significantly and automobiles became accessible to the average person, in the United States to begin with and, years later, in other countries.

The first commercial cell phone, the Motorola DynaTAC 8000X, was registered in the United States in 1983. It cost $3,995 at the time, equivalent to around $11,000 to $12,000 today when adjusted for inflation. The device was bulky, weighed around 2.5 pounds (1 kilogram) and offered around thirty minutes of talk time. Today, smartphones are widespread across all sections of society.

In all of the above cases, it was the initial consumption by wealthy buyers that financed the development and expansion of production. The high prices paid by the rich drove economies of scale, technical advancements, and competition. Products that started out as luxury items later became the standard. The path from elite status object to the mass market is therefore historically the rule and not the exception.

But what are the motives for space tourism today? It is sometimes compared to adventure vacations or extreme sports, although this only really applies to the perception of risk. In the United States, space tourists must first be informed about the potential dangers, including death, injury, physical and psychological damage, and economic loss. They must sign a declaration stating that they have been informed of all these risks, including the fact that there may be other, as yet unknown, risks. Only once they have signed this "written informed consent" form are they allowed to fly. In addition, spaceflights are classed as high-risk activities, which means they are uninsurable.[363] Some critics are unhappy that signing the above waiver is enough to become a passenger on a spaceflight, but I think it's a good thing that, for once, responsibility is placed on the consumer, unlike in other scenarios, where U.S. case law often treats adults like small children who can sue if they scald themselves with a hot coffee, as if they couldn't possibly have expected something like that to happen if they were careless.

But apart from the risk, space tourism in its current form has little in common with an adventure vacation. Space tourists have no influence whatsoever on the course of their journey and are largely passive—at least on short suborbital flights. So what drives people to spend such huge sums of money? The two most frequently cited motives are the experience of weightlessness and the prospect of seeing the Earth from space—both of which sound plausible. Presumably, for some, there is also the prestige associated with being able to afford such an exclusive experience.

The first empirical studies on the motives for space tourism are now available. Some, it has to be said, are of poor methodological quality. In 2024, for instance, a study on the motives for suborbital space travel was published in the *Annals of Tourism Research Empirical Insights*. The study was based on an online survey of 870 participants, of whom only thirty-three (less than 4 percent) had a gross annual income of more than $100,000 a year. This means that higher earners were even less well represented in the survey than they are in the U.S. population as a whole, where one in four earns more than $100,000.[364] In view of the sums that such excursions into space cost, such surveys are of little use.

One qualitative study based on interviews is somewhat more informative, albeit with a very small number of participants. In-depth interviews were conducted with four potential and two actual space tourists. Greg Olsen, who visited the ISS as a space tourist in 2005, said: "People who know me understand that this is going to be a life-changing experience for me. . . . Dennis Tito says he thinks about it every day and when you know you've had an experience like that, you feel special about yourself. My guess is that I will appreciate life and try to do more."[365]

Mark Shuttleworth, who visited the ISS as a space tourist in 2002, explained: "I've always been a geek. I was a bookworm. . . . What's really interesting for me about this space adventure has been how rewarding it is to challenge yourself physically. I never took sport particularly seriously. Of course, now I have to. Some of the physical tests are pretty strenuous."[366] He was attracted by the psychological experience of doing something he was previously very afraid of: "I have moments of great fear when I think about the extreme forces and technological difficulties of manned spaceflight. Walking through that fear has been one of the ways this project has already been personally rewarding."[367]

Preparing for a spaceflight was almost as important as the flight itself for some participants, as one article about the motives of space tourists revealed: "Prospective space tourists have to invest a great deal of time and money, as well as emotional energy, in their undertaking, since they have to endure a long period of training and preparation in order to make the vision a reality. This does not seem to be a problem for most of the participants in this study, and also apparently adds to the attractiveness of the undertaking, in terms of both its novelty value and the level of physical or mental challenge it poses."[368]

Especially in light of the brevity of the spaceflight, the adventure of preparation, which offers the opportunity to connect with like-minded individuals and confront substantial physical and psychological hurdles, probably adds an adventurous dimension to the overall experience for many space tourists today.

Critics, however, vehemently oppose space tourism. For some, these travelers are nothing more than narcissists, suffering from fantasies of omnipotence that warrant psychological analysis. The British sociologist James S. Ormrod wrote a paper on "Pro-Space Activism and Narcissistic Phantasy" in which he blamed capitalism for the proliferation of narcissism, which in turn fuels the yearning for space travel. The desire to experience weightlessness, Ormrod claims, is an expression of the desire to return to the womb.[369] "Capitalism thus serves as the adult's ever attentive mother. This can be witnessed in the emergence of space tourism companies and other private enterprises such as those which 'sell' stars to people, which offer consumers a pre-Oedipal sense of omnipotence."[370]

One common line of criticism is that only the superrich can afford these trips, as stated in one article about the history of space tourism: "We must be aware of the uneven nature of this development, in favoring elites rather than the masses."[371] Following the space race between Bezos, Branson, and the SpaceX missions, widespread criticism emerged, especially from anti-capitalists. The left-wing U.S. politician Bernie Sanders declared, "it's time to tax billionaires [because] here on Earth, in the richest country of the planet, half of our people live paycheck by paycheck, people are struggling to feed themselves, struggling to see a doctor—but hey, the richest guys in the world are off in outer space!" His tweet quickly went viral: "I actually don't think we're angry enough about rich people going to space while the world burns."[372]

Elon Musk, in response to such criticism, has adopted an overly defensive stance—he could have argued far more effectively. "I think," Musk said, "we should spend the vast majority of our resources solving problems on Earth. Like, 99 percent plus of our economy should be dedicated to solving problems on Earth. But I think maybe something like 1 percent, or less than 1 percent, could be applied to extending life beyond Earth."[373]

I have already explained why I find the criticism of space tourism by anti-capitalists such as Sanders to be unconvincing. Firstly, the criticism is unconvincing because the assumption that more money will solve the problems of poverty is fundamentally flawed. Secondly, the criticism is unconvincing because it is true of all new products and services that initially only the rich can afford them. We should be glad that there are superrich people who are willing and able to pay extremely high prices, thereby funding further technological advancements.

But anyone who adopts this argument is also setting themselves up for criticism, because if, at some point, space tourism was to become widely accessible—or at least significantly more affordable than today—another argument will be put forward, namely concern for the environment. Annette Toivonen, who completed her doctorate at the University of Lapland in 2022 on "The emergence of New Space: A grounded theory study of enhancing sustainability in space tourism from the view of Finland," describes the dilemma in which the proponents of space tourism find themselves as follows: "Either way, if the development of space tourism remains only for the minor elite and thus offers 'joyrides' for the ultra-wealthy, or if it extends to a much larger market segment and then exerts a more significant environmental impact, the creation of a new space tourism product, from the environmental point of view, cannot really be justified."[374]

Today, both arguments against space tourism are often combined, as in the polemic against the rich, who have a much bigger "carbon footprint" than the general population. "Since it is also one of the most expensive ultra-luxury experiences, orbital tourism likely constitutes the most extreme existing activity in terms of climate and environmental inequalities," write several authors in an article on "Environmental sustainability of future proposed space activities."[375]

Rocket launches do cause environmental pollution and the current model of producing single use rockets is anything but sustainable. But private companies such as SpaceX or Blue Origin are further ahead than the governmental-led space industry in this respect. Making reusable rockets is undoubtedly an important step toward sustainability. "Despite its relatively short existence," Toivonen writes in her book on *Sustainable Space Tourism*, "the commercial space tourism industry has already presented operational-level sustainability in a way that has not been seen in the traditional governmental-led space industry."[376]

In my opinion, the most significant environmental hazard posed by the space industry is space debris. This encompasses all nonfunctional, human-made objects in Earth's orbit, including defunct satellites, rocket components, and fragments from collisions. A 2025 report mentions 131 million satellite and rocket parts larger than one millimeter.[377]

Due to their extremely high velocities, even the smallest parts have enormous destructive potential. The biggest problem is that this debris does not naturally degrade by itself, but can remain in orbit for years or even centuries. The soo-called "Kessler effect" describes a scenario in which the density of objects in orbit becomes so high that a collision triggers a chain reaction of further collisions, creating more and more debris. Such a development could render entire orbital regions unusable, as the risk of collision for satellites and spacecraft would increase dramatically. International space agencies already have to execute regular evasive maneuvers to safeguard functional satellites or the ISS from debris.

But here too, the primary contributors to this problem are governments and not private space companies. One notable example was the Chinese anti-satellite test in 2007, in which a decommissioned weather satellite was destroyed by a rocket, creating more than 3,000 cataloged pieces of debris. This debris remains scattered across various orbits, posing a threat to other spacecraft.[378]

Another incident was the Russian anti-satellite test in 2021, which destroyed the decommissioned Kosmos-1408 satellite. This test produced over 1,500 identified pieces of debris and thousands of smaller fragments, which are also difficult to track. Following this incident, the ISS crew had to temporarily seek safety in spacecraft due to the heightened risk of collision.[379]

Private companies, including SpaceX, OneWeb, and other satellite constellation operators, have introduced "end-of-life" protocols to responsibly conclude the operational lifespan of their satellites and mitigate space debris. These protocols mandate the controlled deorbiting of satellites once they reach their technical or economic end of life. In low earth orbits (LEOs), this is typically achieved by intentionally lowering the satellite's orbit, allowing it to burn up upon atmospheric reentry within a few years. Some systems have their own engines for this purpose or use atmospheric drag in lower orbits.

With its Starship, SpaceX is also pioneering the concept of suborbital point-to-point transportation. This would revolutionize long-distance travel on Earth by replacing conventional aircraft with a fully reusable spacecraft. Instead of spending hours on flights, passengers could travel distances such as New York to Shanghai in around thirty to forty minutes. This will be possible if the Starship follows a suborbital trajectory, at altitudes of between sixty-two and ninety-three miles, outside Earth's atmosphere, where minimal air resistance allows speeds of up to 17,000 mph.

Such flights could save enormous amounts of time, especially for global business trips or in emergencies. In addition, if powered by methane derived from renewable sources, Starship could offer a more environmentally friendly alternative to long-haul aircraft, potentially emitting less CO_2 per passenger. However, several challenges remain: The development and operation costs are high, despite SpaceX's efforts to reduce expenses through reusability. Launch and landing platforms, known as spaceports, would need to be built worldwide, often offshore, in order to minimize noise and safety risks. The concept would also require modifications to aviation regulations and international agreements, and not every passenger would be able to cope with the high G-forces during takeoff and reentry. Musk first unveiled the concept in 2017, with examples including flights from London to Tokyo or from Los Angeles to Sydney in around thirty minutes. Potential launch sites could be floating platforms off coastal cities such as New York or Singapore.

However, scientists emphasize that the crucial point regarding the relationship between environmental pollution and space tourism lies in distinguishing between short-term pollution and long-term benefits: "Some critics claim that space travel will become a significant environmental burden. However, while superficially correct in the short term, this is the opposite of the truth over the longer term. It would be a dangerous error to prevent the growth of space tourism in order to avoid its initial, minor environmental impact, since this would prevent a range of major benefits in the future, including the supply of low-cost, carbon-neutral SSP [space-based solar power], and other space-based industry."[380]

For several years now, scientists have been developing the concept of space-based solar power—generating solar energy from space. This would entail the construction of large solar power plants in Earth's orbit, which would then transmit the generated energy to Earth via microwave radiation. "Antennas on the ground," says space expert Reichl, "would collect this energy. Vast amounts of energy could be supplied to an area spanning just a few square kilometers. And this energy is entirely CO_2-free during production. Aside from the construction of the rockets (which will be reusable in the future) and the solar cell panels, and the production of the fuels (liquid oxygen and methane, hydrogen or kerosene), no greenhouse gases would be emitted. The subsequent decades of energy production will be emissions-free."[381]

However, this is all still a long way off, and when such concepts are developed, it is likely that solar panels will not be transported from Earth to space but will instead be manufactured directly in space—as explained in the chapter on asteroid mining. In any case, the connection between such forms of energy generation and space tourism is only indirect: "The realisation of such power is dependent on much lower launch costs, which, presumably, only the development of a passenger space travel industry could achieve. The development of orbital tourism could, therefore, provide the key to realising such a source of power economically."[382]

Space tourism remains lightly regulated, largely because it is still in its infancy. With the Commercial Space Launch Amendments Act of 2004, the United States established a transition period for the regulation of private space travel, known as the "learning period." This initially prohibited the Federal Aviation Administration (FAA) from implementing new safety regulations for passengers on commercial spaceflights for eight years—unless there was a serious incident or near-accident.

The aim was to allow the emerging industry the freedom it needed to gain experience and innovate without being hindered by regulatory constraints at an early stage. The moratorium was originally due to end in 2012, but was extended several times so as not to impede the industry's development. The Commercial Space Launch Competitiveness Act of 2015 further extended the deadline to September 30, 2023. In a 2021 report, the FAA itself advocated maintaining the period to continue data collection and support industry growth.

The FAA Reauthorization Act of May 2024 extended the deadline to January 1, 2025. Subsequently, in December 2024, Congress extended the moratorium to January 1, 2028 as part of the National Defense Authorization Act. During this "learning period," the FAA may continue to enforce existing takeoff and reentry regulations, but is prohibited from issuing new passenger safety rules. Instead, as already mentioned, passengers are required to sign a "written informed consent," in which they acknowledge their awareness of the associated risks. The repeated extensions underscore the political will to support commercial space travel as an economically and strategically important growth sector.

I think it would be best to make this moratorium permanent. Clearly, the industry has been able to thrive over the past two decades without the typical government overregulation present in all other areas of life. However, I suspect that this will change as soon as there is a major accident resulting in the loss of life. This, I am afraid, would be followed by very loud calls for massive regulation—and lawmakers would comply, even if the new regulations would have a minimal impact on the likelihood of future accidents.

In these early stages, an elevated accident rate is unavoidable. Around a hundred years ago, flying was also a high-risk venture, given the technical immaturity of aircraft, limited understanding of material fatigue, and rudimentary navigation aids. Nevertheless, early accidents did not lead to government overregulation, but rather they spurred innovation within the industry. Manufacturers improved engines and made advancements in control mechanisms and aircraft structures because demand for commercial flights depended on passengers' confidence in the safety of air travel. Airlines therefore had a vested interest in accident prevention, as each incident directly tarnished their reputation and thus their profitability.

This self-interest led to investment in pilot training, maintenance procedures, and weather forecasting systems long before many of these measures were mandated by law. Technological advances, such as radar, GPS, and modern avionics, drastically reduced human error and hazards arising from poor visibility or navigation. Enhanced materials and redundant systems minimized the impact of technical failures. The companies' systematic evaluation of incidents enabled them to continuously optimize their processes. Consequently, the accident rate steadily declined without a constant increase in regulation serving as the driving force.

Here are some figures: 1929, the year of the stock market's "Great Crash," was also one of the most crash-ridden in aviation history, with twenty-four fatal accidents officially reported. In 1928 and

1929, the overall rate was about one accident for every million miles flown. In today's aviation industry, this accident rate would correspond to around 7,000 fatal accidents per year.[383] In its most recent aviation safety report (five-year average 2020–2024), the International Air Transport Association (IATA) counts one accident every 810,000 flights.[384.]

How important will space tourism be in the future? In general, tourism is one of the most important economic sectors in the world today. In the United States, the tourism industry supports between ten and eighteen million jobs (depending on which definition is used), while in the EU, approximately eleven million people are employed in economic activities directly related to tourism, with the travel and tourism sector as a whole employing almost 25 million.

Much like other new developments, such as the internet, the potential of space tourism has been—and continues to be—greatly overestimated in the short term by its enthusiasts and dramatically underestimated in the long term by its skeptics. In 1908, the Harvard astronomer William Pickering stated: "It is doubtful if aeroplanes will ever cross the ocean, and despite the Wright success they offer little menace to warfare." The idea that in a generation it would be possible to fly to London in a single day, Pickering declared, was "manifestly impossible."[385]

Similarly, in 1915, a professor at the Massachusetts Institute of Technology advised a student who was particularly keen on aircraft engineering to stay in mechanical engineering because "this airplane business will never amount to very much."[386] The student's name was Donald Douglas, and six years later he founded the Douglas Aircraft Company. Under his leadership, the company developed world-famous aircraft, including the DC-3, which revolutionized civil aviation, as well as numerous military aircraft during World

War II. In 1967, Douglas Aircraft merged with McDonnell to form the McDonnell Douglas Corporation, which was later acquired by Boeing.

Who knows how many professors today are telling their students that space tourism to the Moon will never happen? Short suborbital flights are an attraction for some, but unaffordable, unattractive, and too risky for most. But that doesn't mean concepts like space hotels—in space or on the moon—won't have a bright future. If you are young today, you will probably live to see this, and for your children, a short trip to the moon may be as normal as a flight overseas is for you today.

CHAPTER 8

THE SPACE ECONOMY TODAY

Following the Moon landing, Wernher von Braun was asked what the priorities for space travel should be in the years ahead. He replied: "We have to show that spaceflight is useful, and even profitable, for people on Earth, and in fact, space projects should begin to pay for themselves."[387]

While asteroid mining and space tourism, the topics of the two previous chapters, are projects for the future that have no positive impact on peoples' lives today—and from which no (asteroid mining) or hardly any (space tourism) money is currently earned—other major sectors within the space industry are already shaping our lives today, and generating substantial revenues.

The OECD defines the "Space Economy" as "the full range of activities and uses of space resources that create value and benefits for humanity through the exploration, investigation, exploitation, management and use of space."[388] In 2022, according to various estimates, around 400,000 people were working in the Space Economy worldwide.[389] According to these assessments, the Space Economy was worth between $370 and $470 billion in 2021, with satellite navigation and communications accounting for the lion's share.

Looking ahead, the Bank of America projects that the global Space Economy will be worth $1.4 trillion by 2030.[390] Furthermore, a study by the World Economic Forum (WEF) in collaboration with McKinsey forecasts growth from $630 billion in 2023 to $1.8 trillion by 2035,[391] with an average annual growth rate of 9 percent, outpacing the anticipated growth of the global economy. "This growth will largely be built upon space-based and/or enabled technologies such as communications, positioning, navigation and timing; and Earth observation."[392]

The WEF emphasizes the impact of the Space Economy on many companies around the world—an impact that most people are not even aware of: "Entire markets would not exist but for these space technologies, which are considered the 'reach' of the Space Economy. For example, without the combination of satellite signals and chips inside your smartphone, Uber would never have reached such a global scale with its unique ability to connect drivers and riders and provide directions in every city. The reach of the global Space Economy is visible across all industries, representing $300 billion in 2023."[393] The WEF identifies the decreasing costs of launching payloads into orbit and the substantial reduction in satellite prices as key drivers of this development.[394] Here are two examples cited by the WEF:

Satellite-based systems such as GNSS (Global Navigation Satellite Systems) provide high-precision positioning, navigation and timing (PNT) data, essential for the steering, orientation, and safety of autonomous vehicles. The WEF report calls these functions a crucial component of so-called space-enabled activities, without which large-scale autonomous mobility could never work.[395]

As for the Internet of Things (IoT), satellite connectivity means that even remote or poorly connected regions can be integrated into the global data exchange. Space technologies thus enable the precise monitoring, operation, and data collection via networked IoT sensors to deliver enhanced services in areas such as agriculture, logistics, infrastructure, and environmental monitoring. The combination of Earth observation data, GNSS, and satellite communication is described in the report as the key to scalable IoT applications.[396]

But this is just the beginning. Astrophysicist Simonetta Di Pippo, former Human Spaceflight Director at the European Space Agency ESA, illustrates the importance of the space industry: "From agriculture to public health, from education to disaster management, from smart cities to the growing need for water and food for a population that is also growing on a global scale, from tapping alternative energy sources to monitoring the seas: we are progressively becoming a space-based society. . . . Space, therefore, represents today's new frontier and the backbone of tomorrow's economy."[397]

What was once confined to the realm of science fiction is now, partly, reality. As the physicist and science fiction author Arthur C. Clarke wrote back in 1977: "The impact of telecommunications satellites on the entire human race will be at least the same impact as the advent of the telephone in so-called developed societies."[398]

Satellite mega-constellations such as Starlink, OneWeb, Kuiper, and Sat Net will ensure that the third of the world's population that does not currently have access to the internet will soon be connected—with far-reaching economic implications. While military applications are of course important, the Space Economy is today dominated by private enterprise, as Di Pippo explains: "If we analyze the space market from a customer-type perspective, 82 percent are commercial, 9 percent are government or civilian, and 9 percent are defense-related."[399] In recent years, the term New Space has gained traction. The New Space economy is dominated by commercial activity, in contrast to Old Space, which typically refers to activities developed entirely under the auspices of governments.[400]

Few people realize how much our lives already depend on space-based infrastructure. To illustrate this, consider the following thought experiment: If all satellites were to fail, navigation via smartphones and in vehicles would become impossible, weather forecasts would become unavailable, and airports would descend into chaos without GPS-based time and location data. Traffic signals would be desynchronized, leading to chaos on the roads, while supply chains would disintegrate as warehouse logistics and route planning rely on satellite technology.

Even financial transactions would grind to a halt, as precise time signals via satellites are vital for synchronizing debits and credits. For instance, trading on the New York Stock Exchange depends on time stamps accurate to billionths of a second, distributed via GPS; without these signals, automated trading systems could fail. The power grid in many regions also depends on precisely timed GPS control systems and any failure would result in frequency deviations and instability. Telecommunications would also be severely impacted, with mobile phone masts relying on satellite coordination; a failure in this system would result in the collapse of many network services within minutes.

Satellites also play an essential role in modern agriculture, providing vital support for autonomous agricultural machinery through precise positioning data, and in times of drought, GPS-supported, precision agriculture enables efficient water distribution. In remote areas such as parts of Alaska, satellite-based communication is indispensable, serving as the primary connection to the outside world.

Satellite internet in these areas allows access to education, telemedicine, and digital infrastructure. Public transportation systems also rely heavily on satellites: Buses and trains send position data to apps that commuters rely on for real-time information. A disruption in satellite services would paralyze entire passenger information systems.

Similarly, parcel and delivery services, including major companies like DHL and UPS, depend on satellite navigation to optimize their routes. Any failure in this system could result in increased delays and incorrect deliveries. Many of these processes can no longer be executed manually because they rely on machine precision. A breakdown of the satellite infrastructure would cripple countless systems, from ATMs to power load management in substations.

However, the importance of satellites extends far beyond navigation services. Communications satellites are crucial for mobile networks, television, and high-speed broadband internet, particularly in areas lacking terrestrial infrastructure. Without satellites, there would be no global network for internet, telephony, and media transmission.

Weather forecasting is almost inconceivable without satellites. Earth observation satellites provide continuous data on cloud formation, temperature trends, and humidity, which are essential for accurate weather forecasts, disaster management, and early warning systems. In the case of natural disasters such as hurricanes, droughts, or floods, geostationary satellites enable early warnings, speeding up evacuations and rescue operations—and thus demonstrably saving lives.

Of the almost 15,000 active satellites in orbit today, the majority belong to private, profit-oriented companies—with almost 10,000 in SpaceX's Starlink constellation alone.[401] Starlink now has more than six million customers across 140 countries. The company even recently outlined plans to compete with GPS in the near future by providing positioning, navigation, and timing services.[402]

In rural and remote regions, such as parts of Indonesia, Africa, South America, and Oceania, Starlink is often the only broadband option. Any disruption in service could significantly impact education, telemedicine, and e-commerce. In Ukraine, Starlink has been providing critical communications infrastructure for civilians and the military since 2022, playing a key role in the country's defense efforts. This is particularly important as terrestrial communications networks have been targeted by Russian forces, who have also used electronic jamming to disrupt Ukrainian coordination and reconnaissance efforts.

Starlink has also demonstrated its value in disaster response, offering reliable communications for rescue teams during critical incidents such as the 2019/20 Australian wildfires and Hurricane Melissa in Jamaica and the Bahamas in October 2025—where terrestrial networks failed. It has become common practice for Elon Musk to extend Starlink services to threatened regions in the event of natural disasters.

Space technology has also permeated everyday life, as evidenced by the development of a new form of high-precision GPS navigation system. Developed by the U.S. start-up Locus Lock in collaboration with modern satellite companies like Xona Space Systems,[403] this technology enables position data accuracy to within a few centimeters—even in places where conventional GPS systems fail. This innovation has far-reaching, everyday implications, particularly in the realm of autonomous vehicles, delivery robots, and drones. In urban environments with narrow streets and tall buildings, known as "urban canyons," GPS signals are often unreliable due to reflection or obstruction. The technology from Locus Lock combines satellite data with other sensors, including motion detectors, allowing for precise navigation even when traditional receivers fail. This is achieved through sophisticated software that runs on inexpensive, software-defined receivers—in other words, it no longer needs expensive and specialized hardware previously used in military and aviation applications.

For farmers, the precision offered by this GPS technology allows tasks such as field cultivation, fertilizer application, and planting to be carried out with centimeter-level accuracy, all without the need for human drivers—saving resources and increasing efficiency. In the construction industry, machines can be precisely steered to place excavations or foundations. And emergency services and fire departments benefit from reliable location data, even in situations where satellite reception is disrupted, such as during power outages or in tunnels.

Locus Lock is just one of many private space companies that fall under the New Space umbrella today. New companies often emerge because visionaries with groundbreaking ideas are not taken seriously in established institutions and can only implement their ideas outside rigid structures.

One such example is Planet Labs, originally known as Cosmogia, which was founded in 2010 by three former NASA scientists—Chris Boshuizen, Will Marshall, and Robbie Schingler—who had all worked at NASA's prestigious Ames Research Center in Silicon Valley. The director of engineering at the time, Pete Klupar, would sometimes hold up his smartphone at the end of meetings and ask his colleagues to think about what this could mean for space travel. He had realized that the revolution in the consumer goods industry—and above all in smartphones—was opening up opportunities that traditional space companies had no idea about.

While traditional space companies—and many NASA projects—relied on expensive custom-built satellites, Marshall and his colleagues had a different vision: They wanted to build low-cost, mini-satellites inspired by the efficiency and performance of state-of-the-art smartphones like the Nexus One. During their time at the Ames Research Center, they experimented with some initial designs, but soon realized that NASA's bureaucratic structures would severely hamper their ability to pursue their ideas. This prompted them to venture outside the space agency to develop their technologies and led to the birth of Planet Labs in 2010.

Like many start-ups, the company began in a garage in California. Today, it has almost one thousand employees and a market capitalization of $9 billion (as of January 18, 2026).[404] This example disproves the view that government research institutions develop important technologies and private entrepreneurs merely exploit (or steal) them, as suggested by the economist Mariana Mazzucato.[405] In this instance, it was the other way round: The crucial innovations originated in the private sector, but their significance for the space industry had not been recognized by NASA, a government organization. Within NASA's bureaucratic structures, the three scientists would never have been able to transfer technological developments from other sectors into the space industry. And so, a new and innovative company was born.

Planet Labs is a prime example of how the trend toward miniaturization has also revolutionized the satellite business. Satellites used to be big and expensive—sometimes as big as a van or a small school bus and costing up to $1 billion. The founders of Planet Labs recognized the potential of sending a large number of small satellites into space instead of a few very large ones, capitalizing on the rapid advancements in consumer electronics miniaturization—cell phones with cameras were getting smaller and better at the same time.

The first satellites—the company called them Doves—measured 10 x 10 x 30 centimeters and instead of costing $1 billion, they cost less than $1 million to produce.[406] Today, a Dove satellite can be manufactured for well under $100,000. And, unlike their predecessors, these satellites are no longer made to last forever. That would make as little sense as designing a smartphone today to last ten or twenty years. After a few years, new cameras and other components available on the market will be far more advanced, so older satellites would ideally need to be replaced by newer models anyway. The expected operational lifespan of a Dove satellite is therefore only around three to five years.

It takes just over one hundred Dove satellites to photograph every point on Earth every day in such a way that changes become visible. Planet Labs actually operates a constellation of approximately two hundred Dove satellites, which gives the company the redundancy it needs to mitigate the risk of failure and conduct satellite tests in space. Since it was founded in 2010, Planet Labs has launched some 650 satellites into space and, in November 2025 alone, a SpaceX launch carried thirty-six Doves and two high-resolution Pelican reconnaissance satellites into space on a Falcon 9 rocket (Transporter-15 mission).[407]

The small Dove satellites, roughly the size of a shoebox, fly over almost the entire land surface and extensive marine areas of the Earth every day. Planet Labs' "CubeSats" are designed to scan the Earth with a high observation frequency, providing images at 3.7-meter resolution[408] and delivering almost real-time information for various applications, including environmental monitoring and agriculture.[409]

Furthermore, AI-based software detects any changes in the images since the previous day. And for customers who want to know more about specific changes, the acquisition of Terra Bella from Google in 2017 equipped Planet Labs with fifteen larger "SkySat" satellites, which means even more detailed images of any location on Earth are available, including images with resolutions of 50 cm per pixel.[410] Currently, Planet Labs operates more than fifty ground stations around the world to handle data transmissions.

Planet Labs is particularly advanced in the automation of its mission control center. While NASA and ESA control their satellites with large teams of several hundred employees, Planet Labs has automated almost all of the control and coordination of its two hundred satellites. In the event of a satellite malfunction or failure, the system typically restarts automatically, resolving many issues without external intervention. Consequently, a team of only around thirty employees worldwide is required to handle cases that cannot yet be automated.

The company's satellites operate in a sun-synchronous orbit, allowing them to cross every point on Earth's surface every day (and night) at the same local time. This orbit enables consistent observations, as the angle between the sun and Earth's surface remains relatively constant from day to day.

Collectively, the Doves capture several million images per day, which Planet Labs offers to governments, media outlets, and businesses through a subscription service. Journalists often use the photos for research, as exemplified by a BBC report from July 2024 that warned of an impending humanitarian crisis in northern Ethiopia. The BBC had analyzed satellite images showing dried-up reservoirs and fields, even after irrigation.[411]

In another example, in December 2022, the *Washington Post* reported illegal gold mining activities at the summit of a sacred mountain within a protected national park in Venezuela. Despite denials from the socialist government of Maduro—presumably because the politicians had been bribed by the gold miners—the satellite images from Planet Labs, published by the *Washington Post*, provided irrefutable evidence.[412]

There is also great demand for these images in the business world. For example, analyzing the number of cars parked at large retailers at specific times can yield insights into customer numbers and sales performance.

Planet Labs is not the only company operating in this segment. In the Space Economy, this is referred to as the Earth observation or remote sensing sector. Few other sectors of the Space Economy have anywhere near as many start-ups, largely because it's already possible to make money in this field today, in contrast to asteroid mining, which is only a distant prospect.

One problem, however, is that the European Union offers, or increasingly intends to offer, services free of charge that Planet Labs' customers pay for. In the United States, there has been a trend of NASA stepping back from areas where private enterprises can also provide effective services. The situation is different in the European Union. One example is the European Commission's Copernicus program. Formerly known as Global Monitoring for Environment and Security, Copernicus is an Earth observation program jointly established by the European Commission and the ESA in 1998.

The Sentinel-2 satellite constellation, part of the Copernicus program, comprises three identical satellites—Sentinel-2A, Sentinel-2B, and Sentinel-2C—each of which provides images of the Earth's surface at a resolution of approximately ten meters per pixel. Sentinel-2 regularly monitors the Earth's land areas, including vegetation, water, and coastal regions, providing data for climate change and environmental monitoring.

Each satellite is equipped with a multispectral instrument (MSI) capable of capturing images across thirteen spectral bands, from visible light to near-infrared and shortwave infrared. The spatial resolution varies between ten, twenty, and sixty meters, depending on the spectral band. The three satellites are positioned to scan the Earth's surface every three days, ensuring a steady stream of updated data.

The upcoming Sentinel-2 Next Generation program will provide an upgrade on the current Sentinel-2 satellites. The goal of this new generation is to enhance existing capabilities, particularly by enhancing spatial resolution to approximately five meters and increasing the refresh rate of daily image captures. This would create serious competition for Planet Labs, as Sentinel-2 data is publicly and freely available, while Planet Labs offers its data on a commercial basis.

This highlights the damage that can be done when governments spend billions of taxpayer dollars on projects that private companies can execute equally well, if not better, at a fraction of the cost. The government offers services, purportedly "free of charge," but in reality, paid for by taxpayers, thereby engaging in unfair competition with private enterprises—as happened with the space shuttle, which significantly impeded the growth of a private launch industry.

Despite this, the trend toward private companies remains unbroken. Raphael Roettgen, founder of the investment firm E2MC, explains that the Space Economy encompasses a far wider range of categories than most people realize.[413] Besides companies that build rockets or satellites and the "remote sensing" sector, there are companies that specialize in scientific experiments or even production in space, because the weightlessness and vacuum conditions there are impossible to replicate on Earth, or only with great difficulty.

The substantial decrease in launch costs is projected to open up new opportunities for business models within the Space Economy that were previously considered unprofitable. As a result, demand for transportation options is expected to surge, which will in turn drive further cost reductions and foster a positive cycle of decreasing costs and an expanding Space Economy.

SpaceX, for example, frequently launches larger numbers of satellites into space at once as part of its so-called rideshare or bandwagon missions. However, some of these satellites are placed in different orbits than where they are actually needed. These shared flights give smaller satellite providers access to space but little say in exactly where their satellites are released. Recognizing this challenge, some companies have begun to develop "micro-launchers"—smaller rockets that can inject satellites precisely into the desired orbits and at the customer's preferred time. Additionally, new business models have emerged, including "space tugs," modeled on tugboats pulling barges on rivers. Also known as orbital transfer vehicles, these space tugs transport satellites to their target orbits in space. Other companies are working on building the equivalent of orbiting gas stations to provide a means of refueling satellites or rockets that have run out of propellant.

The beauty of capitalism is that, unlike government-run space programs, which tend to focus on a limited number of large, costly projects, it spawns thousands of companies and ideas. While many ventures might not succeed, some, such as SpaceX, have the potential to revolutionize industries in ways that government-owned enterprises never could. Wherever there's a problem, creative companies will emerge and develop solutions and business models, addressing issues like the increasingly critical problem of space debris. Entrepreneurs are simply better at developing innovative solutions than government entities. This doesn't mean that government-run space programs or agencies like NASA are becoming obsolete. They continue to play a role in conducting pure research and serving as clients or partners for private companies.

In this book, we've highlighted the development of SpaceX as an example because the company occupies a special position and is far ahead of the competition. But it is by no means the only company worth considering. Another, Rocket Lab was founded in 2006 by New Zealand engineer Peter Beck. The company was initially headquartered in New Zealand before relocating its main offices to California, where it now operates as a publicly traded company. From the outset, Rocket Lab aimed to expand access to space through smaller, cost-effective launch vehicles.

Its first successful suborbital rocket, Ātea-1, was launched in 2009. The company's first orbital-class launch vehicle, Electron, conducted its maiden flight in January 2018 and remains Rocket Lab's primary launch system. Electron is a two-stage rocket designed for small satellites, with a payload capacity of roughly 300 kilograms to low earth orbit. Its Rutherford engines—largely manufactured using 3D-printed components—use electric pump-fed propulsion, a concept that was unique in the industry at the time.

Alongside Electron, Rocket Lab currently operates the suborbital HASTE (Hypersonic Accelerator Suborbital Test Electron) vehicle, which is used as a test platform for military payloads and experimental vehicles. The company's medium-lift launch vehicle, Neutron, is under development, with a maiden flight planned for mid-2026.

In 2024, Rocket Lab conducted fourteen orbital launches with Electron-13 from its Launch Complex 1 on the Māhia Peninsula in New Zealand and one from its Launch Complex 2 in Virginia in the United States. It also performed two additional launches for the HASTE program on behalf of undisclosed government and defense customers. On December 21, 2025, Rocket Lab reported its twenty-first launch of the year, ranking it—albeit by a substantial margin—second only to SpaceX in terms of total rocket launches.

Rocket Lab, while not yet profitable, is aiming to be so in the next few years—and is experiencing rapid growth. Peter Beck, the company's founder, serves as another compelling counterexample to the economist Mariana Mazzucato's thesis that all innovation ultimately stems from government intervention. Beck initially pursued space technology as a hobby after work and on weekends. Somehow, he managed to convince several venture capitalists to fund his pastime.[414]

"He wanted to spend all day, every day," reports Ashlee Vance, "building rockets but still had to take on contract jobs performing carbon-fiber tests on yachts or retooling some company's gearbox for spare cash. To keep the rocket program advancing, Beck often found himself in a junkyard, crawling around to find pipe fittings or hunks of metal on the cheap."[415]

Beck had hardly any money and even had to take out a second mortgage on his house: "I put our family in pretty serious financial situations. If it all went bad, it would go really bad. My wife was an incredible engineer in her own right but had agreed to stay home with the kids. She gave up a lot and put up with a lot for me to pursue my dream. But it's either you believe in it, or you don't. That's just what you do."[416]

Eventually, against the opposition of his business partner, who walked away from the company as a result, Beck managed to secure some smaller military contracts.[417] Within just one year, he developed a handheld, rocket-launched, mini-drone system for real-time aerial reconnaissance in crisis situations. "It's likely that a military contractor had not worked so fast and at such a reasonable cost since the 1960s," writes Vance.[418]

As with SpaceX, the work ethic at Rocket Lab is quite different from that at traditional companies or government agencies. As one employee recounts: "Peter would give you a job to do, just ask you to look into something or to buy a part. Coming from my government contractor background, I'd figure to get to it in the next two or three days. One of the things was to find someone local who could sew parachutes. Peter asked me to do that and then checked in with me in a couple of hours to see how it was going, and I hadn't gotten anywhere. Not more than half the day had passed, and it turned out that Peter had jumped on the phone and organized it all himself in thirty minutes. He didn't say anything to me or tell me off. But when he wanted something done, he would get it done. He would also devour hard-core rocket textbooks. There was always a stack of them on the shelves. If there was a problem, he'd dive into the books."[419]

While SpaceX's first three launches had ended in flames, as we saw in chapter 4, Rocket Lab's first three had been just about perfect.[420] "No new rocket program had ever started like that," writes Vance. "It had built the best small rocket ever for $100 million and almost done it on time."[421]

Rocket Lab moved its operational headquarters to the United States in order to capitalize on better access to funding, particularly venture capital. The U.S. space industry also offers direct access to key institutional clients like NASA and the Department of Defense. This move was essential for securing contracts with these entities, as they require companies to be registered and based in the United States. Furthermore, the industrial infrastructure and availability of specialized professionals in the United States are superior to those in New Zealand. The relocation was also necessary for the company's subsequent listing on the Nasdaq stock market. In official documents, such as the SEC filing for stock market approval, Rocket Lab makes it clear that the move to the United States was necessary to maintain its long-term competitiveness and be able to hold its own against other U.S. companies like SpaceX.

Despite its growing U.S. presence, Rocket Lab continues to conduct many of its launches from New Zealand, as the Māhia Peninsula offers highly favorable conditions for orbital missions. From this site, sun-synchronous and polar orbits can be reached efficiently without overflying densely populated regions. The surrounding airspace and maritime area are comparatively low-traffic, which significantly simplifies the planning and execution of rocket launches. Although Rocket Lab also maintains a launch facility in the United States (Launch Complex 2 in Virginia), the New Zealand site remains the preferred option for many commercial satellite missions due to its flexibility and operational efficiency.

Anyone who has ever worked in the United States knows that federal and state authorities can be incredibly frustrating with all sorts of regulations. As Ashlee Vance reports, Rocket Lab needed to satisfy approximately 4,300 regulatory requirements in the United States before authorities would give the green light for a launch, compared with only about forty comparable provisions in New Zealand.[422]

In terms of launch costs, companies like Rocket Lab cannot match SpaceX, but the company has carved out a niche that many analysts see as commercially viable. Space expert Eugen Reichl argues that it is misleading to focus solely on cost per kilogram to orbit: "I admire Rocket Lab's high-precision orbital insertions for small, specialized payloads. But in the battle for the lowest transport costs to orbit, Rocket Lab's Electron can't compete. Nevertheless, customers still line up at Peter Beck's launch services company because a range of other factors matter—reliability, the ability to treat even small-payload clients as premium customers, the capability to meet specific orbital requirements, and so on."[423]

Rocket Lab, a publicly traded company, saw its market capitalization rise by 360 percent in 2024, driven by strong financial results, a growing contract pipeline, and the imminent introduction of the partially reusable Neutron launch vehicle. From January 2025 to mid-January 2026, its market capitalization more than tripled to almost $48 billion.

Peter Beck is also interested in Mars. Rocket Lab has submitted a proposal for a Mars Sample Return mission that would retrieve thirty samples collected by NASA's *Perseverance* rover and return them to Earth. The company estimates mission costs at roughly one-third of what NASA had planned. Another initiative is a microsatellite-class Venus probe designed to search for organic compounds in the planet's cloud layers—it would be the first private mission to another planet. Although many launch providers struggle to compete with SpaceX, Rocket Lab demonstrates that there is room for specialized players that serve market segments SpaceX does not occupy.

The Space Economy is growing—and most investments are still being still directed toward satellites and launch services. According to analysts at Space Capital, a total of $358 billion in private capital flowed into the sector between 2009 and the second quarter of 2025, of which $307 billion went to satellites, $40 billion to launches, and $10 billion to emerging industries.[424] Chad Anderson, Space Capital's founder, notes that emerging industries—which he defines as Stations, Lunar, Logistics, and Industrials—have all been widely covered by the media but have accounted for only a small share of total investment to date.[425] However, this will change in the future, particularly due to the sharply reduced launch costs.[426]

Elon Musk and SpaceX have repeatedly emphasized the long-term ambition to expand orbital infrastructure and to enable increasingly sophisticated on-orbit data processing, for example through future generations of Starlink satellites equipped with high-speed laser links. On January 30, 2026, SpaceX submitted a request to the Federal Communications Commission (FCC) for permission to operate up to one million Starlink-style satellites intended to support data centers in orbit. Other actors, such as Google with its research-oriented Project Suncatcher, in cooperation with Planet Labs for prototype satellites, are also exploring concepts for orbital computing.

The starting point of these considerations is the rapid growth in global demand for computing power. Artificial intelligence, cloud computing, streaming services, autonomous systems, and data-intensive research are driving the electricity consumption of data centers sharply upward. Already today, data centers rank among the largest industrial consumers of electricity, and demand driven by AI applications is growing significantly faster than in other segments of the IT sector. At the same time, traditional locations are reaching their limits: Power grids are overloaded, permitting processes take years, suitable land is scarce, cooling water is becoming a political issue, and environmental groups increasingly protest against the ecological footprint of large facilities. In addition, there is a geopolitical dimension: Data infrastructure is increasingly regarded as critical infrastructure, and dependence on specific countries, networks, or energy sources is perceived as a strategic risk.

From these pressures arise several potential advantages of space-based data centers. The most important concerns energy. In low earth orbit, and particularly in sun-synchronous orbital configurations, solar energy is available for a very large fraction of the orbital cycle. While periods of Earth shadow still occur, atmospheric losses, cloud cover, and most seasonal fluctuations are eliminated. As a result, power generation in orbit can be more stable and achieve higher effective utilization than on Earth. Large solar arrays could, in principle, provide power over long periods without local environmental impact and without CO_2 emissions during operation, although emissions from launches and manufacturing must be considered separately.

A second central aspect concerns cooling. In the vacuum of space, waste heat can only be dissipated through thermal radiation, without the use of large quantities of water or energy-intensive cooling systems as on Earth. However, radiation is physically the least efficient mode of heat transfer, which makes very large deployable radiator structures necessary. These structures are heavy, costly to launch, and potentially vulnerable to micrometeoroids. Despite the extremely low ambient temperature of space, cooling is therefore more challenging than often assumed, since convection is impossible. Passive radiators are nevertheless operationally reliable, and hybrid approaches, such as liquid cooling within modules combined with external radiative heat rejection, are considered technically realistic.

Third, many of the classic location- and construction-related conflicts associated with terrestrial data centers do not apply in space. While space-based projects remain subject to regulatory frameworks, including spectrum allocation, launch licensing, and international liability regimes, the lengthy approval processes related to land use, water consumption, and local environmental impact largely disappear.

Fourth, proximity to space-based systems matters. Satellites generate ever-larger volumes of data, for example for Earth observation, communications, and navigation. Processing parts of this data directly in orbit can reduce transmission volumes to Earth, relieve pressure on radio frequencies, and improve overall system efficiency. Finally, resilience is often cited as an additional benefit. Space-based data centers would be independent of regional power outages, natural disasters, or local political interventions and could serve as a redundant complement to terrestrial systems. They also alter the security profile of digital infrastructure: While risks from space debris exist, certain physical threats and attack scenarios are more difficult to implement than for ground-based facilities.

These potential advantages are counterbalanced by substantial technical and economic challenges. For many years, launch costs were considered the primary obstacle. The dramatic reduction in launch costs over the past decade, and the prospect of further declines, particularly with reusable systems such as Starship, have moved space-based data centers from a purely theoretical concept into the realm of serious discussion. Other challenges remain unresolved. Unlike on Earth, defective components cannot simply be replaced, and in-orbit repairs are complex and costly.

Another major issue is radiation exposure in space, which can damage electronics and shorten the lifespan of chips through effects such as single-event upsets, total ionizing dose, and displacement damage. Mitigation strategies include radiation-hardened components, additional shielding, error-correcting codes, and system-level redundancy and overprovisioning. Tests conducted in recent years suggest that even commercial high-performance processors can operate for several years in low earth orbit with moderate protection, particularly for specialized workloads with limited fault-tolerance requirements.

Projects such as Starcloud, which launched an experimental satellite equipped with a high-performance GPU in late 2025, and

Google's Project Suncatcher, announced as a research initiative with prototype missions planned for the coming years, demonstrate that space-based computing systems are technically feasible and already enable early experimental applications. Commercial scaling on a large industrial level, however, remains a matter for the future.

The success of SpaceX—especially in reusable launch vehicles and with its Starlink constellation—has inspired countless entrepreneurs worldwide. It appears that we are witnessing a Bannister effect with SpaceX: When Roger Bannister ran a mile in under four minutes for the first time (3:59.4 minutes) in Oxford on May 6, 1954, his achievement was recognized globally as a milestone in athletics. What followed, however, was arguably even more remarkable. In the years immediately after Bannister's record-breaking run, the so-called four-minute barrier was repeatedly shattered by numerous other athletes. Just 46 days later, the Australian runner John Landy completed a mile in Finland in 3:57.9 minutes—considerably faster. Over the next few years, the record continued to be broken.

This development is often cited in research as an example of the influence of psychological barriers. The four-minute mark, long considered "impossible," clearly had a primarily psychological component. As soon as Bannister showed that it was possible, other elite runners no longer regarded this barrier as insurmountable—and their performances improved dramatically. Thus, a feat previously considered "impossible" is no longer seen as such once it has been achieved. In this way, the Bannister effect serves as a motivating factor for countless others.

The benefits of space exploration extend far beyond the areas I have presented in this chapter. The philosopher of science Gonzalo Munévar has presented a comprehensive "Philosophy of Space Exploration" under the title *The Dimming of Starlight*. His argument:

Only by understanding other planets can we gaining a deeper understanding of our own planet (just like you cannot understand a country without comparing it to others). "We cannot vary the global conditions of our planet at will. But we can look at other worlds in which those variations occur naturally and see how other factors are correlated with them."[427]

One example of the practical application of comparative planetology is that when NASA scientists found fluorine and chlorine compounds in the atmosphere of Venus, they investigated the chemistry of those molecules and determined the rate constants of their chemical reactions. Those rate constants were later used by Sherwood Roland and Mario Molina to discover that chlorofluorocarbons (CFCs) destroy ozone in the presence of high ultraviolet radiation.[428]

Understanding other worlds, Munévar shows, advances the way we understand our own planet. Only space research allows us to understand the development of the solar system and the Earth and opens "opportunities to test our ideas about the Earth—the solar system serves as a natural laboratory."[429]

Research, Munévar emphasizes, is inherently unpredictable, often leading to unexpected discoveries and applications. One example of this phenomenon is the work of astrophysicists Anil Pradhan and Sultana Nahar from The Ohio State University, who were investigating the composition of stars by analyzing the flow of radiation through them. This research led directly to groundbreaking insights into cancer treatment.[430] This case illustrates how basic research in one field can unexpectedly pave the way for breakthroughs in an entirely different domain. Such cross-disciplinary connections typically only become evident in retrospect and cannot be deliberately planned.

The impact of space technology is far-reaching and a list of all the things developed for space that have found their ways into our daily lives would fill a book. From heat-resistant ceramics to lightweight composites, the materials developed for space programs have found their way into sporting goods, cars, and airplanes. Sensor technologies initially created for space missions now benefit medical imaging and diagnostic techniques. Even wireless tools and water filtration systems can trace their origins back to developments optimized for use in space.

This phenomenon is known as the "spillover effect," where technologies, processes, or materials originally designed for space applications are transferred to other industries and everyday products, such as in medicine, communications, and materials science. Many examples of spillover effects come from the traditional space industry with its governmental stakeholders. Lars Hornruf and Daniel Vrankar from the University of Dresden, however, analyzed 35,696 space-related patent applications and granted patents from the last ten years and concluded that New Space companies have a greater potential for such effects and are far more adept at transferring space innovations to other industries than large aerospace companies.[431]

One of the reasons the authors cite is, "the increased willingness of new space firms to integrate commercial off-the-shelf components into their systems, which are substantially cheaper than so-called graded components. This behaviour increases spillover potential, as space innovations becomes less complex and costly, making them easier to implement in terrestrial applications."[432]

A master's thesis on "Aerospace technologies and civil spillovers" by Gianluca Librera, published in December 2022, points in a similar direction. The study is interesting because, above all, it analyzes start-ups and small and medium-sized aerospace firms. Government space agencies and large firms or multinationals operating in the aerospace industry are not included in the analysis.[433]

The analysis is based upon a patent dataset. The researcher examined how often and by whom aerospace companies' patents are cited by other patents. This allows conclusions to be drawn about how technologies spread between companies (technological spillovers). Additionally, European start-ups and small to medium-sized enterprises were identified using data from the dealroom.co platform to detect the firms belonging to the space sector that have filed patents. Statistical analyses demonstrated that knowledge transfer from the aerospace industry occurs more frequently to other industries than within the industry itself. "The patents cited only by citing patents not classified as 'space' (18.43 percent) are almost seven times the number of patents cited just by other 'space' patents (2.67 percent)."[434]

This indicates that aerospace technologies—particularly in the *New Space* sector—are highly adaptable and increasingly find application across a wide range of civilian domains. As private *New Space* companies continue to gain prominence relative to legacy government space agencies and their major contractors, even greater cross-sector spillover effects can be expected in the future than in the past.

In various discussions, I have sometimes been told that the world managed perfectly well before the advent of satellites and the other technologies described in this chapter. By the same logic, an argument could be made that before the invention of electric light and the printing press, people could read books by candlelight—books that were copied by hand—and before telephones and email, people could write letters. All of these objections are valid, and yet I am nevertheless grateful that I can read by electric light and communicate instantly by telephone.

CHAPTER 9

CHINA'S PRIVATE SPACE INDUSTRY

In 2003, China completed its first manned spaceflight mission with Shenzhou 5, becoming the third nation, after the United States and the Soviet Union, to independently launch humans into space. Ten years later, in 2013, China's Chang'e 3 lander touched down on the lunar surface carrying the Yutu rover. Then, in 2019, Chang'e 4 became the first spacecraft to land on the far side of the Moon—a unique undertaking, since direct transmissions between the Earth and the far side of the Moon are impossible. To achieve communications with Chang'e 4, China deployed a communication relay satellite in a halo orbit with a clear view of both the lunar landing site and the Earth. The mission investigated the geological environment of the South Pole Aitken (SPA) basin and conducted biological experiments, including the germination of cotton seeds.[435] In follow-up missions, Chang'e 5 returned lunar rocks to Earth in 2020, and Chang'e 6 retrieved nearly two kilograms of lunar material in June 2024.[436]

China's future space missions are set to be even more ambitious. The *Tianwen-3* mission aims to collect samples from Mars and return them to Earth for the first time in history. The Mars lander will be equipped with a drill to extract samples from a depth of approximately three meters at the landing site. A helicopter, also being sent to Mars, will explore geological formations, which will then be sampled by the rover. If all goes according to plan, the mission will begin in 2028, and the first rock samples will arrive back on Earth in 2031. The United States has long planned a similar mission, but at the moment it looks as if China will be the first to achieve this milestone.[437]

In 2021, China began construction of the Tiangong space station, which has housed Chinese astronauts regularly since June 2022. China launches crewed missions to the station approximately every six months to rotate crews and continue scientific experiments. Like the ISS, the space station is permanently manned, albeit with a smaller team, typically consisting of three astronauts. The Chinese "talkonaut" corps, comprising forty-eight members, is comparable in size to the astronaut corps of the United States.[438]

In terms of rocket launches, China is the second-largest spacefaring nation after the United States. However, significant problems exist—for example, China's Starlink alternative, the "Qianfan" ("Thousand Sails") project, which aims to deploy 15,000 satellites by 2030, is currently a long way behind schedule, due in part to limited rocket resources.[439] While China's primary focus remains on state-run space programs, a notable private space sector has been emerging since the mid- to late 2010s, and that is the focus of this chapter.

But before we examine China's space program, it is important to address some misconceptions about the Chinese economic system, which is often misunderstood in the West. China is frequently labeled a "communist country," in which the state plans the entire economy. This characterization is just as inaccurate as describing the United States as a purely capitalist country. Yes, during the Mao era there was no private property, and socialist experiments had a devastating impact on the country. A stark example is Mao's so-called Great Leap Forward, which resulted in the deaths of approximately 45 million people between 1958 and 1962.[440] As recently as 1981, just a few years after Mao's death, more than 88 percent of the Chinese population was living in extreme poverty.[441]

In the 1980s and 1990s, Deng Xiaoping initiated economic reforms that allowed for private property. The 14th National Congress of the Communist Party of China in October 1992 officially declared the establishment of a market economy as the goal of these reforms. Although the planned economy was not abolished, the proportion of prices of raw materials, transportation services, and capital goods that were set by the state decreased dramatically. Furthermore, reforms of state-owned enterprises (SOEs) commenced, permitting private individuals and foreign investors to become shareholders. The privatization process quickly gained momentum, with some companies being listed on the stock exchange.

The economist Zhang Weiying of Peking University, arguably the most astute analyst of the Chinese economy and a personal contributor to its development, argues against the notion that China's extraordinary success is a result of the state's major role. This misinterpretation is widespread in the West, but it has also now become increasingly prevalent in China, where politicians and academics believe the explanation for the country's success lies in a unique "Chinese model." Zhang disagrees: "The advocates of the China model are wrong because they mistake 'in spite of' for 'because of.'

China has grown fast *not because of*, but *in spite of* the unlimited government and the large inefficient state sector."[442] In fact, market economy and privatization are the driving forces behind China's enormous economic growth. Zhang analyzed data from different regions in the Middle Kingdom. He came to the following conclusion: "The more the market-oriented reform a province had done, the higher economic growth it had achieved, and laggards in marketization reform are also laggards in economic growth."[443]

At a World Economic Forum congress in June 2025, Roger Hu and Carol Liao from the Boston Consulting Group reported: "China's private companies contribute 60 percent of the country's gross domestic product (GDP), 70 percent of technological innovation, 80 percent of urban jobs, and 90 percent of the total number of enterprises. Most of these are family-owned and their numbers have grown so quickly that they now account for 67 percent of the companies listed on the Shanghai and Shenzhen stock exchanges."[444]

Whether these rather symbolic figures, which have been cited repeatedly for many years, are still accurate today is difficult to say. Since Xi Jinping assumed leadership in China about a decade ago, the trajectory of development has shifted, with the state playing an increasingly prominent role again—undoubtedly a significant factor in the slowdown of China's economic dynamism. However, as we shall see, this trend does not extend to the space industry, where while state-owned enterprises continue to dominate, an increasing number of private companies have entered the sector during Xi Jinping's tenure.

The number one player in China's space sector is the state-owned China Aerospace Science and Technology Corporation (CASC). CASC is involved in "virtually every state-initiated space program, from the development of launch vehicles to the development and construction of communications satellites, the manned space program, and lunar and planetary programs."[445] Its most important subsidiaries are the China Aerospace Science and Industry Corporation (CASIC), which primarily focuses on military projects; the China Academy of Launch Vehicle Technology (CALT), which develops launch vehicles; the China Academy of Space Technology (CAST); and the Shanghai Academy of Spaceflight Technology (SAST).[446]

In the mid-2000s, shortly after Xi Jinping took office, a series of reforms commenced, outlined in a strategic document known as "Document 60." These reforms allowed private investors to enter the space sector, leading to the emergence of commercial space companies. Brian Harvey writes in his book *China in Space*: "The development of new commercial launchers was very much a response to the success of SpaceX. At first, SpaceX was not taken seriously, but by the mid-2010s the state adopted an 'if you can't beat them, join them' approach and licensed the development of private rocket companies in China."[447] SpaceX's influence on the Chinese space industry should not be underestimated. "The American SpaceX company was even praised by Li Yanhong, a delegate to the National People's Congress. Elon Musk visited Beijing to promote his Tesla cars, but it was his film of the Falcon rocket that attracted the most attention."[448]

Blaine Curcio, one of the world's leading experts on the Chinese space industry, reports: "Up until 2014, 100 percent of space activity [in China] was done by the state. But even after that, it was not clear for a while if there was going to be any business for these commercial companies. It wasn't clear if they were going to be allowed to access launch sites—it wasn't clear what was going to be allowed to them."[449]

While the state sought to harness the entrepreneurial innovation evident in the United States' private space sector, it was not yet prepared to cede control to these new companies, many of which originated as spin-offs from the large state-owned enterprises mentioned above. "Most of the commercial space companies that were founded toward the beginning still had some pretty strong links to the state," says Curcio. "It was typically engineers from CASC that would group together to work on a problem like reusable launch vehicles. They could do it with CASC, but because CASC moves very slowly, they would make a commercial company instead."[450]

Since 2015, an increasing number of entrepreneurs and engineers have entered the burgeoning private space market, attracted by the business opportunities and the promise of greater technological autonomy compared to state-owned enterprises. By 2022, there were 430 private space companies,[451] and by 2024, the combined value of the 100 most valuable space companies in China was estimated at approximately 733 billion yuan, equivalent to about $100 billion.[452]

However, the boundaries between private and state-owned companies can be fluid, as Brian Harvey writes: "The dividing lines between public and private companies, generally clear and strict elsewhere, are often blurred in China, with some being a mixture of state, private and local government."[453] Often, the state or provinces in China hold significant stakes in private companies, and to this day, the private companies almost always have to purchase their rocket engines from state-owned defense contractors.[454]

The Chinese government supports private space companies with subsidies, tax breaks, access to launch sites, and so on. However, this close relationship with the government also creates disadvantages and risks, as Gareth Morgan Thomas writes in his book *The Chinese Space Dynasty*: "The relationship between the state and private companies in China's space sector is complex and sometimes fraught with challenges. While the government provides substantial support, it also imposes stringent regulations that can stifle innovation and operational flexibility. The dependence on state support can make private companies vulnerable to shifts in policy and government priorities."[455] This is true, but similar observations could also be made about American private space companies.

Nevertheless, the growth of private space companies in China is remarkable. "Analysts attribute the causes of this rapid expansion to two key factors. First, the large state-run aerospace firms struggled to meet the surging market demand for cost-effective commercial solutions to space, such as smaller satellites and reusable rockets. SOEs' bureaucratic inertia and risk-averse work culture of older, conservative development teams could not keep with up national goals or markets for faster innovation and better space systems. Second, the rise of Elon Musk's SpaceX, with its reusable Falcon 9 rockets and Starlink mega constellation served as both an inspiration and warning. China recognized the need for a fast-paced and risk-taking private sector innovation if it hoped to remain competitive in the space domain that is critical to both military and civil interests."[456]

The last point is of great importance. As is well known, the Chinese revere the teachings of the philosopher Confucius, who once said: "There are three ways to attain wisdom: First, by reflection, which is noblest; Second, by imitation, which is easiest; and third by experience, which is the bitterest." While the West likes to be "original," China has long excelled at learning from international successes. And, of course, the achievements of private companies like SpaceX have certainly not gone unnoticed in China.

"Over the last four or five years, there's been a larger trend towards more openness to commercial companies doing bigger things in China," explained Blaine Curcio, space and satcom industry consultant at satellite consultancy firm Novaspace and one of the world's leading experts on China's commercial space sector, in 2024. "I think a lot of that is being driven by the rapid progress of SpaceX and Starlink. They see Starlink every week going into a new country with a new product. Soon there's not going to be any countries left."[457]

China, Curcio added, recognized the strategic importance of Starlink in the Russia-Ukraine conflict, and wants a similar asset. But the Chinese government is also aware of the limits of its state-run enterprises, which have so far been responsible for delivering the nation's space program. "They see that the state-run companies are innovating in their slow state-owned way, so they are getting more interested in the idea of commercial companies becoming an important part of the space sector just as they are in the west," Curcio said.[458]

China's ambition to equal the United States is a big motivator for the country's willingness to embrace the commercial approach, said Robert Lincoln Hines, an assistant professor in the Sam Nunn School of International Affairs at the Georgia Institute of Technology whose research focuses on China's security and foreign policy. "The state wants to make sure that it maintains its competitiveness with the United States and the international market more broadly in space and thinks that having some private capital within the space industry could incentivize innovation, so that it can be more competitive on the international market," he said.[459]

The fascination surrounding SpaceX, particularly Starship, and the Starlink satellite system in China, is enormous. "Every Starship launch, every Starlink launch, and frankly every word Elon says is covered with interest," Blain Curcio told the global space industry news service SpaceNews in March 2025. "If you line up all the big milestones that SpaceX has had, and what China is doing, there's a bit of a lag, but there's a clear mimicking."[460]

But can China replicate the success of companies like SpaceX? First, China's leaders would never tolerate a figure as powerful and independent as Elon Musk. The tale of Jack Ma, founder of Alibaba and Ant Group, serves as a case in point. Long regarded as a symbol of China's economic ascent and the success of private entrepreneurship, Ma was a global celebrity as one of China's richest men, with an estimated net worth exceeding $50 billion. In October 2020, Ma publicly criticized China's financial regulations as stifling innovation, a move widely construed as an affront to the government. Shortly thereafter, the government halted the planned IPO of Ant Group, which, at over $30 billion, would have the biggest IPO in human history. Ma subsequently disappeared from public view for several months, sparking international speculation about his fate. Simultaneously, the Chinese government launched a sweeping regulatory crackdown on major technology firms, particularly Alibaba. The company was fined a record $2.8 billion and forced to restructure. Ma was effectively stripped of power, losing all influence over the management of his companies and ceding numerous control rights. Ma's fate has also served as a warning to other private entrepreneurs in China, making it more inconceivable than ever before that an Elon Musk–like figure will emerge there.

Tereza Pultarova quotes space expert Hines, who believes that despite the rivalry with SpaceX, many in China are not convinced that commercial spaceflight will truly take off in their country. Ambitious projects, such as the development of giant reusable rockets to rival SpaceX's may encounter reluctance from potential investors unwilling to assume the high levels of risk embraced by the American billionaires Musk and Bezos.[461] "The really prominent space companies in the United States have in part succeeded because people like Elon Musk are willing to watch $3 billion or $5 billion explode on a launch pad," said Hines. "That is a stomach for risk that is very uncommon, and the question is whether any of the private entrepreneurs in China are willing to put their capital on the line to the same extent."[462]

So, which of the hundreds of private Chinese space companies currently have the highest profiles? China's space industry, according to Curcio, "has impressive breadth and a very deep bench. This is in contrast to the U.S. space industry today, in which the vast majority of rockets and satellites being launched are coming from one mega-company with an impressive degree of vertical integration."[463]

Curcio highlights the structural differences between the American and Chinese space industries: "I always say that if we compare the most impressive commercial launch company in the west with the most impressive in China, the west is far ahead. And if we compare #2 in both places, probably the same. But if we compare #5 in the west with #5 in China, it gets closer. And #10, in China is undeniably in the lead. And #20 probably the same. And #30 . . . you get the idea. The same applies for satellites, laser communication terminals, thrusters, and the rest of the supply chain."[464]

And new companies are constantly emerging in China's space industry. At a recent conference, a representative from a Chinese state-owned space company estimated that there are now at least thirty-five start-ups in the launch sector alone.[465] Here, therefore, follow only a few of the better-known names as examples:

Founded in 2015, LandSpace was one of the first private space companies in China. In 2023, its best-known rocket, Zhuque-2, became the first methane-powered rocket to successfully reach orbit—a world first. That same year, the U.S. companies SpaceX and Relativity Space also attempted launches of methane-powered rockets—Starship and Terran 1, respectively—but both failed. SpaceX's Starship exploded shortly after liftoff, and Terran 1 did not reach orbit. Methane is widely considered an efficient, cost-effective, and forward-looking propellant option, particularly for reusable launch systems and future interplanetary missions. As we have seen, Starship also relies on liquid methane and liquid oxygen, a combination known as methalox, in part due to its potential for production on Mars.

i-Space, founded in 2016, was the first Chinese private company to launch a rocket, Hyperbola-1, into orbit in 2019. The company is now developing more advanced rockets, including Hyperbola-2, which incorporate reusable architectures.

Galactic Energy, founded in 2018, had completed twenty-three launches of its Ceres-1 rocket by January 2026, 21 of which were successful.[466] The company also plans to develop reusable, liquid-propellant rockets such as Pallas-1 in the medium term, and aims to move into asteroid mining in the longer term.

CAS Space, often classed as a "commercial entity" despite its affiliation with the Chinese Academy of Sciences, is developing the Kinetica family of rockets and is planning both suborbital and orbital missions. It intends to offer tourist flights starting in 2027.

Deep Blue Aerospace was founded in 2016 and is developing a reusable orbital rocket, Nebula-1. The company also plans to enter the space-tourism market by offering suborbital passenger flights starting in 2027. The first two tickets were sold during a widely publicized live stream event in October 2024. This again illustrates China's interest in learning from American companies, as the Nebula-1 rocket bears a strong resemblance to the New Shepard system developed by Jeff Bezos' Blue Origin. The crew capsule closely matches New Shepard in both size and appearance. And Blue Origin and Deep Blue Aerospace are pursuing the same suborbital tourism business model: a reusable booster with a passenger capsule mounted on top, providing passengers with a few minutes of weightlessness before returning via a parachute or retro-propulsion system.[467]

China is also making progress in the area of rocket reusability. On May 30, 2025, the private company Sepoch successfully conducted a test flight of China's first reusable liquid oxygen-methane rocket, followed by a recovery at sea. The test was conducted at the Oriental Aerospace Port in Haiyang, in East China's Shandong province. The company claimed on social media that the test integrated multiple rocket technologies, including liquid oxygen-methane propulsion. The rocket's stainless-steel structure enabled a soft landing at sea, and all phases of the mission proceeded smoothly, culminating in a successful postlanding recovery.[468]

These companies are financed with a mix of private capital and state funding, often supplemented by investments from provincial and municipal governments. Such support is often tied to the condition that companies establish factories or other infrastructure within a particular city or province.[469] This may have contributed to the increasingly fragmented landscape of production facilities now scattered across China. According to Curcio, the number of such facilities has increased tenfold, from approximately ten in 2015 to around one hundred in 2025.[470]

This highlights the negative impact of political influence, a phenomenon also evident in the United States, where contracts are frequently awarded based on the number of jobs created within a given state. U.S. companies are often drawn to particular states by generous incentive packages and tax breaks, and they later express their appreciation to the governor, senator, or member of Congress who facilitated these arrangements through substantial campaign contributions. Regarding China, Curcio warns against excessive regional fragmentation within the space industry. "Should cities such as Chishou, Deyang, Tai'an, and Shaoxing be angling to become hives of commercial launch activity? Probably not. And do the industrial development funds of such cities have the skill sets to do real due diligence on launch company investment pitches? Also . . . probably not."[471]

In China, the production of rockets and satellites, along with satellite-based communications services, are among the largest space technology sectors. These are heavily subsidized by the state, with public sources accounting for 40 to 50 percent of funding. State capital also dominates in the operations and services sectors, accounting for 50 to 70 percent. One reason for this could be that the relatively low margins offered by these business models, which make them less appealing to private investors. Future technologies are also largely state-owned, as they are closely linked to public research institutions. At the same time, private investors are increasingly channeling funds into areas such as remote sensing, navigation services, and the production of components for rockets and satellites.[472]

From the outside, it is difficult to assess the true state of China's private space sector. The fact is, the Chinese state plays a more dominant role in its space economy than the federal government does in the United States. Whether China's space sector will evolve along a path similar to that of the United States over the past two decades hinges upon whether ideological or pragmatic approaches prevail. Arguments in favor of strengthening the private sector include the successes of SpaceX and other U.S. companies, which have demonstrated to the Chinese that private enterprises can often be far more efficient and innovative than large state-owned entities. Conversely, arguments against such a shift include China's reluctance to tolerate highly successful entrepreneurs like Elon Musk gaining prominence. It is nevertheless noteworthy that, contrary to the general trend under Xi Jinping—namely, a stronger role for the state—the private space industry has emerged and continues to develop under his leadership.

In their book *Red Moon Rising: How America Will Beat China on the Final Frontier*, Greg Autry and Peter Navarro argue: "We will not beat China at socialism by running a centrally planned, governmental space race. We were able to do that with the Soviets in Space Race 1.0 because our nation was the 'workshop to the world' and we could out-manufacture and outspend anyone. If we are going to win Space Race 2.0, we must out-innovate China and retain our capital and IP for use in growing our commercial sector."[473] They further assert: "Winning the second space race is all about the private sector. We won't beat China in a competition of large governmental programs; commercial space is America's best weapon."[474]

That is undoubtedly true. Essentially, the question is which of these two countries grants more freedom to private space exploration. Currently, that's the United States, but China has taken note and has been steadily shifting away from a reliance solely on state-run space programs for more than a decade. And the Chinese shouldn't be underestimated: Even if they are currently still learning from SpaceX, Blue Origin, and other U.S. companies, this may not remain the case for much longer. In the automotive industry, for example, the Chinese have already proven that they can transition from copying Western products to innovating for themselves, overtaking Western companies.

For example: Geely was founded in 1986 as a manufacturer of refrigerator parts before venturing into the automotive sector in

1997, becoming China's first privately financed car company. Under the leadership of the private entrepreneur Li Shufu, the company's chairman and majority shareholder, Geely became the first Chinese car brand to sell more than ten million vehicles worldwide at the end of October 2020. By 2024, Geely had sold more than 17 million vehicles and had been the leading Chinese automaker by sales volume for five years in a row.[475]

Parallel to its automotive strategy, Geely is also investing in space exploration. As early as 2018, the group founded a subsidiary, Geespace, to develop a low earth orbit satellite constellation. As of November 2025, Geespace had a fleet of sixty-four satellites in orbit,[476] with plans to expand the constellation to 240 satellites. These satellites will not only provide high-precision positioning data, but also Earth observation imaging with a resolution of one to five meters. The ultimate goal is to connect vehicles—particularly autonomous models—via navigation, real-time connectivity, and AI-driven remote sensing, thereby giving autonomous mobility a major boost. Terminals and chips for satellite communication are already being mass-produced and integrated into vehicles across Geely's brands.

At the same time, Geespace is expanding its international activities and has formed partnerships with network operators in more than twenty countries, including a joint venture with Malaysia's Altel Group to expand coverage in Southeast Asia. This makes Geely one of the few automakers worldwide to develop its own satellite infrastructure, positioning itself, in some respects, as a competitor to aerospace companies like SpaceX.

The West must not underestimate China's space program. The common perception of a simple dichotomy between a "communist planned economy" in China and a "capitalist economy" in the United States is not only a gross oversimplification—it is also wrong.

CHAPTER 10

CONQUERING SPACE WON'T WORK WITHOUT PRIVATE PROPERTY

Let me take you on a thought experiment: Imagine you start an asteroid mining company. It's tough at first. You know that profits are uncertain and, if they do ever materialize, it will be in five years at the earliest, maybe even ten. It takes a long time to find investors willing to trust you because several similar ventures with almost identical business models have already gone bust and the investment horizon is far too long for many. Nevertheless, you persevere, and you manage to convince a few investors.

You hire astronomers to identify asteroids with valuable metals such as platinum. You send unmanned probes to several near-Earth asteroids to collect samples. The first few turn out to be of little value. The costs are too high and outweigh any potential profits.

Then comes the breakthrough: You find a suitable asteroid, and you succeed in bringing platinum group metals all the way back to

Earth. But then a group of countries file a lawsuit against you. They are all signatories to the so-called Moon Agreement of 1979, formally known as the "Agreement Governing the Activities of States in the Moon and other Celestial Bodies," which came into effect in 1984. The treaty applies not only to the Moon, but to all celestial bodies, including asteroids.

Some of these countries are now demanding that you hand over a significant share of your profits to people who have never had anything to do with asteroid mining—many of whom couldn't even say exactly what an asteroid is, let alone launch rockets into orbit. They invoke Article IV, Paragraph 1, and Article XI, Paragraphs 1 and 3, of the Moon Agreement, which state: "The exploration and use of the Moon shall be the province of all mankind and shall be carried out for the benefit and in the interest of all countries, irrespective of their degree of economic and scientific development. Due regard shall be paid to the interests of present and future generations as well as to the need to promote higher standards of living and conditions of economic and social progress and development in accordance with the Charter of the United Nations."[477] So much for Article IV.

Article XI states: "The Moon and its natural resources are the common heritage of mankind. Neither the surface nor the subsurface of the Moon, nor any part thereof or natural resources in place, shall become property of any State, international intergovernmental or non-governmental organization, national organization or non-governmental entity or of any natural person."[478]

The Moon Agreement, since signed by twenty-two countries and ratified by eighteen, was developed to work in the same way as Seabed law, and a delegate from Sri Lanka explained in simple terms what the intention was: "If you touch the nodules [valuable mineral deposits on the seabed] at the bottom of the sea, you touch my property. If you take them away, you take away my property."[479]

The plaintiffs now demand their "rightful share" of your profits, arguing that the platinum belongs not to you, but to all of humanity—and especially to developing countries, many of which are not yet engaged in space exploration. Should the plaintiffs win, neither you nor any other company would ever be able to attract investors again. Asteroid Mining would be dead and buried after this very first attempt.

But let's pause here for a moment—I will discuss the Moon Agreement, its history, and its relevance in greater detail later. For now, suffice it to say: Investors can remain confident, as a lawsuit based on a treaty signed by only a small number of countries is unlikely to succeed. Before we move on, consider a second thought experiment:

It is the year 2075. The first settlements on Mars have been successfully established, housing several thousand people and many more robots. The settlers are busy building underground houses and domed communities, a hospital, and other facilities. But there is a problem: Unlike on Earth, there is no private property ownership on Mars. The settlers have repeatedly attempted to establish private property rights and even create a digital land registry, but legal experts keep putting a stop to their efforts. This time, the problem is not the 1984 Moon Agreement, but the 1967 Outer Space Treaty (OST), Article II of which states: "Outer space, including the Moon and other celestial bodies is not subject to national appropriation by claim of sovereignty, by means of use or occupation, or by any other means."[480]

And Article I of the Treaty states that the use of outer space "shall be carried out for the benefit and in the interest of all countries, irrespective of their degree of economic or scientific development, and shall be the province of all mankind."[481]

In this case, the argument that the Moon Agreement has limited signatories cannot be made, as 117 countries are parties to the Outer Space Treaty—including all major spacefaring nations, such as the United States, China, and Russia—and another 22 are signatories.[482]

The plaintiffs assert that Article II of the Treaty prohibits not only the national appropriation of celestial bodies, but also any form of private property ownership. They further contend that Article I should be interpreted as mandating the equitable sharing of proceeds from resource extraction among all nations—an interpretation that has sometimes been put forward in the real world.[483]

I will examine whether they are correct in their reading of the OST later. For now, let us assume their interpretation is correct. There are no property rights: Nations cannot own land (which they could potentially lease), nor can private entities, since any form of appropriation by states or private individuals is prohibited. Could such a society exist, let alone be economically successful? Of course not.

Most successful countries permit the private ownership of land. Where this right does not exist, it is at least possible to lease land from the state for fifty to ninety-nine years, which can also be resold—as in Vietnam and China, for example, or (regarding the ninety-nine years) in Singapore. North Korea is the only country where even that is not possible. Property ownership on Mars would therefore be most comparable to that in North Korea, a country plagued by widespread poverty where people go hungry after every poor harvest.[484] Do you believe a settlement on Mars could be successful if its economic model mirrored that of North Korea?

What would possibly motivate people and companies to develop and settle on Mars if they cannot acquire private property or are forced to hand over a portion of their profits from its cultivation and development? A society on Mars, facing environmental conditions more challenging than any on Earth (even Antarctica is far more hospitable to humans because it has oxygen, normal gravity, and atmospheric pressure), would require a maximally efficient economic system (at least as market-oriented as Singapore's), not a maximally inefficient one.

There is much to suggest that Mars would remain in a similar state to Antarctica for this reason alone. The Antarctic Treaty, which was signed in 1959 and entered into force in 1961, imposes stringent restrictions on the extraction of raw materials and both the acquisition and ownership of land in the region surrounding the South Pole. Article IV of the Antarctic Treaty suspends all existing territorial claims and prohibits new claims or the expansion of existing rights for as long as the treaty is in force. Consequently, no sovereignty claims are recognized, and no state can claim Antarctica as its territory or exercise sovereign rights there.[485]

The prohibition on raw material extraction is explicitly defined in the "Environmental Protocol to the Antarctic Treaty" (also known as the "Madrid Protocol"), which was ratified in 1991 and entered into force in 1998. Article 7 of the Protocol states: "Any activity relating to mineral resources, other than scientific research, shall be prohibited."[486] This effectively bans the commercial extraction of mineral resources, including metals, fossil fuels, and other raw materials, across Antarctica. This prohibition was initially established for fifty years, until 2048, after which it can be reviewed, although any changes require the consensus of all contracting parties.

But now, more on the history of the Outer Space Treaty and the Moon Agreement, and the discussions about their interpretation. The late 1950s marked the beginning of the space race between the United States and the Soviet Union, with the Soviet Union initially taking the lead. At that time, it was unclear who would win and both superpowers sought to prevent a space arms race, particularly involving nuclear weapons. In December 1963, the United Nations adopted an important resolution, which would later served as the basis of the Outer Space Treaty. The resolution stipulated that outer space, including the Moon and other celestial bodies, could only be used for peaceful purposes. Military activities, and in particular the deployment of weapons of mass destruction in space, were thus prohibited. The exploration and use of outer space, the United Nations insisted, should be carried out for the benefit and in the interests of all nations on Earth, irrespective of their degree of economic or scientific development.

The resolution stipulated that outer space and celestial bodies are not subject to national appropriation by claim of sovereignty, by means of use or occupation, or by any other means. It further called for international cooperation in the exploration and use of outer space. States were obligated to bear responsibility for any and all activities in outer space, whether carried out by governmental agencies or by nongovernmental entities. The resolution formed the basis for the 1967 "Outer Space Treaty," which enshrined these principles in legally binding international law for the first time.

To put this into perspective, it is important to understand that at that time, the exploration and use of outer space were not approached from the standpoint of economic exploitation and the extraction of raw materials or energy. Instead, outer space was regarded as a "fragile jewel to be observed and studied."[487]

This perspective was also reflected in the 1967 OST, officially the "Treaty on Principles Governing the Activities of States in the Exploration and Use of Outer Space, including the Moon and Other Celestial Bodies," which established the primary international legal framework for all activities conducted in outer space.

Article I of the OST upholds the freedom of all states to explore and use outer space, while Article IV places specific limitations on military activities. It explicitly prohibits the deployment of nuclear weapons and other weapons of mass destruction in Earth's orbit, on the Moon, or on other celestial bodies. Furthermore, the development of military bases, installations, or fortifications on celestial bodies, as well as the testing of any form of weaponry there, is also prohibited. However, the OST does not prohibit all forms of military use of outer space, leaving a certain gray area, particularly regarding the positioning of conventional weapons in Earth's orbit and the use of military satellites for communication or reconnaissance purposes, which several states exploit in practice.

In addition to these provisions relevant to security policy, the OST also sets out the fundamental principles of international responsibility: Article VI stipulates that states are responsible for all national activities in outer space, regardless of whether these are carried out by governmental agencies or by nongovernmental entities. Article VII, in turn, establishes a state's liability for damage caused by objects launched into outer space from its territory or facility, irrespective of whether the damage occurs on the Earth, in air space, or in outer space.

Article II is used to settle the question of property rights, as mentioned earlier. Legal scholars have put forward differing viewpoints over its interpretation. The primary area of contention revolves around the "prohibition of appropriation" and whether this applies only to the *states* expressly mentioned in the Treaty or also to *private individuals and companies* not mentioned in Article II.

In "Outer Space: Problems of Law and Policy," Glenn H. Reynold and Robert P. Merges contend that: "Being prevented from claiming sovereignty and exclusive property rights located in the space environment for themselves, it will be argued that States are also prohibited from granting quasi-sovereign and exclusive property rights over such areas and resources to those natural and juridical persons which are subject to national jurisdiction and which are created through international agreements."[488]

Those who argue that appropriation by private individuals is prohibited also assert that private ownership would infringe upon the freedoms of access, exploration, and use guaranteed to all states in Article I.[489] If private property rights existed, argues space law expert Marcus Schladebach, this would impinge upon the freedoms granted to all states, and the legal status of outer space as a "global commons" beyond territorial sovereignty would be jeopardized. If national appropriation is prohibited, Schladebach maintains, then private appropriation must be prohibited all the more.

Furthermore, he claims, states must be prevented from entering into private legal forms to circumvent the prohibition on national appropriation. According to Schladebach, Article VI of the OST is clear that such prohibitions apply equally to nongovernmental entities, for which the state nevertheless bears full responsibility. This concept, namely that the state always "authorizes" and "supervises" its private space companies, is for Schladebach also a decisive factor in any interpretation of Article II.[490]

In contrast, other legal scholars argue the exact opposite: National sovereignty stops where outer space begins, which means that national appropriation of the Moon, other planets, and asteroids is forbidden—but not the private ownership of celestial bodies.[491] This interpretation rests on the legal doctrine *expressio unius est exclusio alterius*: The explicit mention of one thing implies the exclusion of others. If, for example, a statute or treaty expressly mentions one or more things of a class, it is only reasonable to conclude that others of the same class remain unrestricted by that statute or treaty. This principle serves to interpret legal norms and supports the assumption that things not mentioned were excluded by deliberate choice, not inadvertence.[492]

Other legal scholars, however, argue that while it is not forbidden for private individuals or companies "to claim property . . . it is a crime for a nation to recognize such a claim publicly." This interpretation, however, confuses the terms "recognize" and "confer," as the legal scholars Alan Wasser and Douglas Jobes explain: "'To recognize' means to 'acknowledge the existence, validity, or legality of' or 'accepts, acquiesces to, decides not to contest.' In contrast, 'to confer' means to 'grant (a title, degree, benefit, or right).'"

They go on: "If a nation claims the right to confer, give, or grant title to lunar land, it could be violating the ban on national appropriation. But if a settlement is established and the settlers claim private ownership of land around their settlement, and a dozen of Earth's nations recognize the settlers' claim, it is not reasonable to say that all dozen nations are trying to appropriate the land and thus are violating the Outer Space Treaty."[493]

Still others argue based on various provisions of the Outer Space Treaty, which, for example, hold states liable if a rocket launched by a private company from their territory crashes in another country. Furthermore, they claim, states are obligated to authorize and supervise the activities of private companies located within their territories. From this, they conclude that, even if not explicitly stated, the Outer Space Treaty prohibits private appropriation.

Wasser and Jobes rightly object to this: "But the treaty clearly does not contain any language explicitly saying that states may not authorize their citizens to do anything that they themselves cannot do, contrary to what some authors appear to assume. The treaty does not say that what is prohibited to states is therefore prohibited to private entities nor that what is prohibited to the regulator is therefore always prohibited to the regulated. A baseball coach gives 'authorization and continuing supervision' to his players. Does the fact that the coach is not allowed to run onto the field to catch a fly ball mean the players he supervises cannot either?"[494]

The wording of Article I of the Outer Space Treaty states that outer space is "the province of all mankind" and that celestial bodies should be "free for exploration and use by all States without discrimination of any kind on basis of equality" and "there shall be free access to all areas of celestial bodies."[495] Critics argue that this turns space into a "public good whose owner is everybody and nobody."[496]

The legal scholar Schladebach, on the other hand, argues that the extraction of mineral resources on the Moon and other celestial bodies is prohibited by Article II of the Outer Space Treaty, even for private companies. His reasoning, however, is unconvincing. Among other things, he argues: "How anyone could read approval for the extraction of lunar resources into the Outer Space Treaty, drafted in the mid-1960s, without knowing at the time whether humans would ever land on the Moon, seems incomprehensible."[497] One could just as easily say: "How anyone could read a prohibition on the extraction of lunar resources into the Outer Space Treaty, drafted in the mid-1960s, without knowing at the time whether humans would ever land on the Moon, seems incomprehensible."

Indeed, presumably no one back then could have imagined that private companies would fund missions to the Moon or asteroids to conduct mining operations. Why prohibit something no one has considered?

The only relatively undisputed point is that the OST is ambiguous in several areas, particularly concerning private property ownership and the role of private companies. The legal scholar Ezra Reinstein articulated this ambiguity in the interpretation of the Outer Space Treaty in a scholarly article, stating: "We have a choice between interpreting the OST as truly hostile to private development, or merely vague."[498]

This ambiguity stems from the priorities of the United States and the Soviet Union at the time of the OST's drafting. Their primary focus was on preventing an arms race in space and the deployment of weapons of mass destruction there. The concept of private space companies, including business models that would allow private companies to engage in asteroid mining, was a distant prospect in 1967. Consequently, the issue of private ownership was a peripheral matter of little importance for both nations. However, the Soviet Union would have preferred to prohibit private companies in space during the treaty negotiations, a proposal the United States rejected.[499] So, they did what national governments sometimes do in joint international declarations or treaties: The language was kept vague and omitted contentious issues, which is what leads to the difficulties interpreting it today.

Many years later, some countries felt the need to clarify the issues that had remained ambiguous at the time. We have seen that—unlike Article II of the Outer Space Treaty—Article XI, paragraph 3 of the Moon Agreement explicitly prohibits private ownership in outer space. This is often cited as evidence that Article II of the OST prohibits appropriation by nations but not by private individuals: "The very fact that the framers of the Moon Agreement felt the need to write a new specific ban on private property indicates that they did not feel the earlier Outer Space Treaty had already accomplished such a prohibition."[500]

The Moon Agreement, which not only explicitly prohibits private property but, according to some interpretations, even demands an egalitarian distribution of the profits generated by private users, breathes the spirit of socialism. "At its heart," says space expert Rand Simberg, "the Moon Treaty was redistributionist in nature, taking from those who were willing to take risk and invest capital in developing new resources and giving to those who did not."[501]

Most countries refused to sign it, albeit for different reasons. This raises the question: If the Moon Agreement merely precisely regulated exactly what the Outer Space Treaty intended—that is, if the Moon Agreement simply served as a concretization of the Outer Space Treaty—then why did 121 of the 139 states that signed the Outer Space Treaty refuse to sign the Moon Agreement? Oddly enough, this question is rarely asked.

What can be done when a treaty, in this case the OST, is open to so many different interpretations? Firstly, it could be renegotiated. For example, in 1999, Ezra Reinstein argued in a legal journal: "What is needed is an amendment to the Outer Space Treaty, one that both clarifies and expands property rights in space."[502] First, Reinstein insists, it must be clarified in the amendment that the formulation, "'for the benefit . . . of all countries' is a moral exhortation and not a loophole through which the United Nations can dispossess a private party of his site."[503]

The right to use a chosen site, Reinstein suggested, should only be limited by environmental regulations and domestic law in the developer's country. He also stressed the importance of enshrining the "right to exclude" in any amendment.[504]

The "right to exclude" is a fundamental concept in property law. It is considered one of the most important coercive measures among the bundle of rights in property law. This right enables owners to control access to their land, buildings, or other possessions and, at the very least, to take legal action against unauthorized persons.

The fact that Reinstein considered all these clarifications important enough to include in an amendment once again underscores the vagueness and legal uncertainty of the Outer Space Treaty. Perhaps there was an opportunity to modify or renegotiate the treaty in the late 1980s or early 1990s, a period marked by the global decline of socialism and the introduction of private property rights in countries like China and Russia.

At that time, there was a widespread belief in the superiority of a society based on private property and free enterprise over socialist ideologies. If such an opportunity existed, it was not seized.

However, it is unlikely that a consensus on a new treaty could be reached today. The geopolitical landscape has shifted: Xi Jinping's China is no longer Deng Xiaoping's China, and Putin's Russia is no longer Gorbachev's Russia.

In 2020, the head of the Russian space agency made clear: "We will not, in any case, accept any attempts to privatize the Moon. It is illegal, it runs counter to international law."[505]

On the other hand, an opportunity may emerge from the current situation, which is similar in one respect to the mid-1960s, when the USSR and the United States originally negotiated the Outer Space Treaty: At that time, it wasn't clear who would win the space race. If I'm in a position where I'm not sure whether I will come out on top or not, I am more likely to be open to fair regulations. Today, the outcome of the race to Mars between the United States and China remains unclear. Although the United States, with SpaceX, appears to be leading, China's capabilities should not be underestimated, as I established in chapter 9. And even if Xi Jinping isn't Deng Xiaoping, China does not fundamentally oppose private property.

The United States should at least try to leverage this potentially limited window of opportunity to engage in discussions with China about amending the Outer Space Treaty to allow private property and clarify the ambiguities of Article II. The success of any such negotiations would be uncertain, but it would be worth a try.

However, this is just one potential avenue and probably not a likely one. In recent years, the United States has adopted a different strategy, one that is certainly a step in the right direction: If a treaty is subject to varied interpretations by different countries, it increasingly becomes the responsibility of national legislators to interpret the treaty. And this is what has happened in recent years, especially in the United States, where the Republican senators Ted Cruz and Marco Rubio championed a "Commercial Space Launch Competitiveness Act," signed into law by President Barack Obama in 2015. This gave all U.S. citizens and companies the right, "to possess, own, transport, use, and sell [any] asteroid resource or space resource obtained in accordance with applicable law, including the international obligations of the United States."[506]

The company Planetary Resources strongly advocated for this clarification to clarify how the United States interprets the Outer Space Treaty. Similar laws were subsequently passed in Luxembourg, the United Arab Emirates (UAE), and Japan. However, Russia, China, and many developing countries opposed this interpretation of the OST.[507]

In April 2020, Donald Trump went a step further than Obama by signing an "Executive Order on Encouraging International Support for the Recovery and Use of Space Resources," explicitly rejecting the notion of space as a "global commons." The Moon Agreement, which the United States had never signed anyway, was also explicitly dismissed.[508]

This executive order was followed just one month later by NASA announcing the "Artemis Accords," officially "Principles for cooperation in the civil exploration and use of the Moon, Mars, Comets and Asteroids for Peaceful Purposes." One key element of the accords is the introduction of what are referred to as "safety zones"—geographically defined areas around a station or activity on the Moon or another celestial body. These zones are not intended to contradict territorial appropriation, but rather to serve the practical implementation of the principle of peaceful use and to prevent potential conflicts arising from activities that are too close to one another. Section 10 of the accords explicitly states: "The Signatories affirm that the extraction of space resources does not inherently constitute national appropriation under Article II of the Outer Space Treaty, and that contracts and other legal instruments relating to space resources should be consistent with that Treaty."[509]

The accords were initially signed by eight nations: the United States, Australia, Canada, Italy, Japan, Luxembourg, the United Arab Emirates, and the United Kingdom. Since then, the number of signatories has expanded to sixty. Russia and China, however, have rejected the accords and criticize their interpretation of the Outer Space Treaty. China, in particular, is concerned that the United States may be using the Artemis Accords to assert sovereignty over lunar resources.[510]

While the Artemis Accords were a step forward, the language suggesting that space mining does "not inherently" violate Article II—which prohibits "national appropriation"—has again created ambiguity, as space law scholars Michael Byers and Aaron Boley point out in their book *Who Owns Outer Space?*: "Is Space mining sometimes 'national appropriation' and sometimes not? Was Space mining originally not 'national appropriation,' but capable of becoming understood as 'national appropriation' as understandings and interests change? Can a term such as 'national appropriation,' which has no 'ordinary meaning' because it is not used outside the Outer Space Treaty, 'inherently' mean anything?"[511]

The "safety zones" mentioned in the Artemis Accords could serve as a pragmatic and indirect means of establishing de facto private ownership. It is only logical that areas where, for example, raw materials are extracted must be appropriately secured and demarcated. However, such extraction activities could last for decades, blurring the line between temporary use and de facto occupation of the terrain.[512]

In any case, opposition to the accords soon emerged and, in August 2020, 140 experts penned an "International Open Letter on Space Mining" to the president of the United Nations General Assembly, then Volkan Bozkır of Turkey. The letter called for the swift negotiation of a multilateral agreement on space resource exploration.

This led to the formation of a "Working Group on Space Resources" in May 2021, consisting of representatives from Austria, Belgium, the Czech Republic, Finland, Germany, Greece, Slovakia, and Spain, which emphasized the importance of using resources "for the benefit and in the interests of all countries . . . irrespective of their degree of economic or scientific development."[513]

Austria and Belgium, both signatories of the Moon Agreement, were again actively involved in these discussions—and again the idea was that "all countries" should benefit, including those without any expertise in space mining or space exploration in general.

For a believer in the power of the market, it is anyway obvious that all of humanity will automatically benefit from space mining, but that's clearly not what is meant here. Rather, a recurring, typically socialist notion is hidden behind all this: that it is the duty of more advanced companies and countries engaged in such activities to share the fruits of their efforts with others.

So, what should the goal be? Well, without full private ownership, there will be no colonization of Mars, and the economic exploitation of asteroids and the Moon will also be difficult. And who should own land on Mars or the Moon first? Various people, including the American Dennis Hope, have made a business out of selling land on the Moon without being able to clearly explain why they consider themselves entitled to do so. Hope argued that the Outer Space Treaty prohibited nations—but not private individuals—from acquiring celestial bodies. What he could never explain, however, was the legal basis for him to declare every celestial body in the solar system as his own.

A Spanish woman even started selling shares in the sun because Hope had generously refrained from claiming it, contenting himself with all the other celestial bodies in the solar system. The local Spanish land registry office even issued her a certificate of ownership.[514] This is, of course, absurd. Someone might as well declare tomorrow that they own the entire universe and start selling off planets, perhaps even "cheaper by the dozen."

Space law expert Virgiliu Pop has written an entire book with the delightfully evocative title *Unreal Estate: The Men who Sold the*

Moon, in which he establishes that from the 1890s to the present day, numerous individuals have claimed ownership of the Moon or other celestial bodies—and found plenty of buyers for their properties.[515] It should be clear, he writes, "that claims to extraterrestrial real estate unsubstantiated by physical acts of possession are not valid means of acquiring ownership. . . . A mere claim is not tantamount with ownership—or, in plain language, claiming does not mean owning."[516]

But apparently, the desire to own property in space is so strong for some that it overrides their reason. At the same time, it's also good news to see so much interest in plots on the Moon. Six million customers are said to have bought from Hope, including Tom Cruise, Nicole Kidman, and John Travolta. Hope even claims that two former presidents of the United States have purchased certificates of ownership for plots on the Moon, without, however, naming them.[517]

These transactions offer hope that there will be sufficient interest when plots of land do actually become available on the Moon or Mars, although they will be much more expensive than Hope's offerings. Incidentally, Hope has himself become a multimillionaire thanks to his certificates.

So, who should have the right to acquire property in space? My answer: Those who have the financial means to get there, develop, and use the land. For instance, if SpaceX succeeds in reaching Mars and starts to build permanent settlements on the Red Planet, then the ownership of land should go to SpaceX first. Not of the entire planet, of course, but of a manageable area, for example the size of Singapore. The surface area of Mars is 200,000 times that of Singapore, so SpaceX would initially only own 0.0005 percent of Mars. That would be enough to develop multiple settlements, but not so many that others would no longer have a chance.

SpaceX could fund its flight and development costs by listing the land on Mars in a real estate investment trust (REIT). The price would then be determined by the market. Most people would buy shares not to live there themselves, but in the hope of value appreciation. As an incentive for people to settle on and develop Mars, colonists could be offered stocks at a preferential price as a "golden hello" once they reach Mars and spend at least five years there.

When the Chinese reach Mars, they may assert their claim to plots of land on the planet according to the idea outlined above, and this would probably remain the property of the Chinese state. But it is likely that they would adopt the customary practice in China of selling hereditary building rights for a period of fifty to seventy years.

However, we would need to tread carefully: If the Chinese were the first to reach Mars, it is possible they would declare the entire planet as belonging to China. This would constitute a clear violation of Article II of the Outer Space Treaty, but according to Article XVI, the treaty can be terminated with one year's notice. The potential response from the United States in such a scenario raises concerns—would it lead to war? Or sanctions? Therefore, it would be prudent to negotiate a new space treaty before it is clear who wins the race to Mars. Moreover, competition for premium real estate on Mars, such as in areas with anticipated water sources or suitable lava caves for settlement, would likely ensue. It could be similar on the Moon.

As far as asteroids are concerned, at least in the case of smaller celestial bodies, only those who are capable of mining and extracting resources such as water and platinum should be owners. The best option would probably be to list the entire asteroid on the stock exchange as a REIT, which would fund mining operations and enable shareholders to receive dividends from the extraction of raw materials. Even before a single dollar was earned, or a single penny in dividends could be paid out, this would allow a market for trading in such stocks to develop.

While this is all just a thought experiment, it illustrates the direction things would need to go in. Whether these ideas are realistic or if different concepts will prevail is something no one can know today. One thing, however, is absolutely clear: As long as there is no legal certainty for investors, they will not invest. "If exploitation of outer space's bounty is our goal, we must establish a space property legal system that creates both incentives and predictability. Space development is a highly risky endeavor, as well as mind-bogglingly expensive. Who would expend the effort in developing a space colony, if they were not certain of the project's legality? Valuable projects—energy collection, mining, and colonization—are by no means inevitable. If the law of outer space rejects such uses, or even makes their legality uncertain, it is unlikely that the necessary technology would ever be created."[518]

To return to the starting point of this chapter: Who will invest in asteroid mining if, as some envision, they are expected to share the profits with others who have taken no risk themselves but simply want to profit as free riders?

The bestselling authors Kelly and Zach Weinersmith, already quoted in chapter 5, cite the Moon Agreement as a model in their book *A City on Mars*: "The Moon Agreement would have set up the solar system as a particularly communal form of *res communis*, known in international law as 'common heritage of mankind' or just 'CHM' . . . a commons collectively owned by all of humanity. If the Moon were under a CHM framework and you wanted to use Moon water, you would have to compensate *all of humanity* by some means."[519]

They envision something like a large international planning authority to regulate how the proceeds are distributed: "An international regime to oversee exploitation . . . a big entity established by states that were parties to the Moon Agreement looking over things, and in particular making sure developing nations got a fair cut."[520] This international regime would regulate both "where people are allowed to set up shop and what they're allowed to do with the local resources once they get there. It wouldn't be dynamic, it wouldn't be like a science fiction novel, and frankly it would be very slow and bureaucratic and boring."[521] In this vision, there would be no room for private companies, no room for the formation of fair prices in the market, no room for entrepreneurial ingenuity. It is space socialism, doomed from the start.

On Earth, only capitalism has worked, but for a few decades, socialist systems without private property did manage to survive, albeit at the cost of widespread poverty, authoritarian rule, and famine. Eighty percent of the people who died in famines during the twentieth century perished in Stalin's socialist Soviet Union and Mao's China.[522] In space, where the challenges are immensely greater and living conditions infinitely harsher, socialism would be doomed from the outset.

Even "democratic socialism", as proposed by Chris Hajduk in his book *Terraforming Mars*, won't work on Mars. He writes: "It is expected that within this design the colonists will naturally evolve into some form of Direct Democratic Socialism."[523] No, democratic socialism, even with private property rights, has never worked anywhere in the world and certainly won't work on Mars.

The question of whether establishing private property rights on celestial bodies will be possible, and whether Martians will opt for a capitalist or socialist system, is essentially the same as asking whether colonization is possible at all. Even if Martian settlers were initially prohibited from establishing property ownership and introducing a capitalist system, they would inevitably have to do so sooner rather than later anyway, because they simply couldn't exist on the Red Planet otherwise. I'm sure the Martian settlers won't care all that much about a space treaty signed over one hundred years ago and would establish private property rights. And how would Earth's governments respond? Would they wage war against the Martian settlers, or kill them with sanctions?

AFTERWORD

The Moon landing of 1969 was only possible thanks to an extraordinary national effort in which money was no object. The United States felt compelled to embark on its Moon Shot to counter the perception of Soviet scientific superiority in the wake of the Sputnik launch of 1957. At that time, 58 percent of people in the United Kingdom believed that the Soviet Union was ahead in scientific development, with only 20 percent believing the United States was in the lead. In France, the ratio was 49 percent to the USSR and only 11 percent to the United States; in Italy, it was 37 to 23 percent. Only in West Germany did the United States hold a slight perceived lead, with a ratio of 32 percent to 36 percent.[524]

The Soviet Union's perceived technical—or even economic—superiority over the United States was objectively false, but the communists, as so often, were far superior to the capitalists in public relations and understood the symbolic power of launching the first satellite into space. The success of their propaganda strategy was further cemented when Yuri Gagarin became the first human in space on April 12, 1961. According to American newspapers, it was: "a psychological victory of the first magnitude," and "new evidence of Soviet superiority," that "cost the nation heavily in prestige," "marred the political and psychological image of the country abroad," and meant "neutral nations may come to believe the wave of the future is Russian."[525]

On the same day, President John F. Kennedy faced intense scrutiny at a press conference, fielding twenty questions about space exploration and the Soviets. While Kennedy initially deflected these questions with generalities, he was eventually confronted with a challenging question that would dominate the headlines the following day: "Mr. President, a member of Congress said today that he was tired of seeing the United States second to Russia in the space field,"[526] one reporter began. "What is the prospect that we will catch up with Russia and perhaps surpass Russia in the field?"[527]

The unnamed Congressman was James G. Fulton from Dormont, Pennsylvania, a suburb of Pittsburgh. He had indeed said: "I'm darned well tired of coming in second-best all the time."[528] And when NASA administrators James Webb and Hugh Dryden appeared before the House Science and Astronautics Committee the following day, the influential Fulton told Webb: "Tell me how much money you need, and this committee will authorize all you need."[529] Another Congressman declared: "I want to see our country mobilized to a wartime basis, because we are at war."[530] That was the prevailing sentiment at the time. No matter the cost, they wanted to show that the United States was indeed better.

Four months later, in September 1962, John F. Kennedy delivered his famous Moon Shot Speech at Rice University. But in a private conversation with then-NASA chief James Webb Kennedy admitted: "I'm not that interested in space." For him, the race to the Moon served a singular purpose: "To beat [the Soviet Union] and demonstrate that starting behind it, as we did by a couple of years, by God we passed them. . . . I think it would be a helluva thing for us."[531]

His vice president and eventual successor, Lyndon B. Johnson, put it this way: "One can predict with confidence that failure to master space means being second best in the crucial arena of our Cold War world. In the eyes of the world, first in space means first, period; second in space is second in everything."[532]

The Apollo program cost over $300 billion in today's dollars, and at its peak, had 400,000 people and 20,000 companies working on it. They were all focused on a single goal—and that goal was achieved. But, as former NASA chief economist Alexander MacDonald stressed: "The Apollo program should not be seen as the classic model of American space exploration, but rather as an anomaly. From a long-run historical perspective, this 'Apollo Anomaly' represented an exciting new paradigm for American space exploration, but ultimately a short-lived and ephemeral one."[533]

Well, characterizing the Moon landing as an "anomaly" is valid only when you consider the short time frame in which this monumental achievement was accomplished. But the Moon landing itself stands as one of the greatest endeavors in human history. The problem, however, was that once the objective of demonstrating superiority over the Soviet Union was achieved, it was back to the daily grind of politics.

As I have already shown, the subsequent space shuttle program was a complete failure. The program failed to achieve the often-conflicting objectives set by its government sponsors, leading to skyrocketing costs and, ultimately, each launch devoured over $1 billion. We have seen that for political decision-makers, the primary concern was the number of jobs created in key electoral districts, not the program's success.

In a frequently quoted 1992 speech "On Self-Licking Ice Cream Cones," S. Pete Worden, then working for the Strategic Defense

Initiative Organization at the Department of Defense and later director of NASA's Ames Research Center, took aim at the politicians: "Since NASA effectively works for the most porkish part of Congress, it is not surprising that their programs are designed to maximize and perpetuate jobs programs in key Congressional districts. The Space Shuttle-Space Station is an outrageous example. Almost two thirds of NASA's budget is tied up in this self-licking program. The Shuttle is an unbelievably costly way to get to space at $1 billion a pop. The Space Station is a silly design. Yet, this Station is designed so that it can only be built by the Shuttle and the Shuttle is the only way to construct the Station. Furthermore, the Shuttle has to be 'improved' to support the Station with a new solid rocket motor, which is to be built, *you guessed it*, in the district of the Chairman of the House Appropriations Committee. Since there are tens of thousands of jobs tied up in these programs and most of NASA's budget as well, there is not only no money to get out of this endless do-loop. There are positive political pressures to make sure that we don't get out. Witness the fact that not even $175 million could be found out of NASA's $14 billion budget to develop a new, cost-effective launch system."[534]

I haven't addressed the Space Station *Freedom* program in this book. It was never actually built and was superseded by the International Space Station (ISS), but the station program does indeed confirm the above findings. In his analysis *The Space Station Decision*, Howard E. McCurdy elaborated on the political mechanisms at play in the decision-making process. He described the principle of "incremental politics," whereby decisions were not made based on a coherent, long-term strategy, but through small, politically feasible steps. This approach was geared less toward efficiency and clear objectives than it was to securing short-term majorities in Congress.

From the standpoint of public choice theory, the rationale becomes evident: Political actors do not perform as impartial planners interested in substantive goals, but rather as stakeholders responding to incentives. For members of Congress, this primarily means securing reelection through visible achievements in their constituencies, such as industrial contracts and job creation. For institutions like NASA, it means stabilizing or expanding budgets, mitigating risks, and securing political backing.

The Space Station *Freedom* exemplified this logic from its inception. Contracts were awarded in a way that benefited as many states as possible, which made the program politically unassailable, but at the same time escalated costs and compromised technical coherence. The station and the space shuttle mutually reinforced each other's justification for existence, creating a self-sustaining cycle that critics described as a job creation program for the aerospace industry. As with the space shuttle, the actual purpose of this space station was entirely unclear. A space station program should never be pursued as an end in itself but only as a means to achieve certain long-term goals, otherwise it ends up—as was also the case with the space shuttle—as a compromise between a several interest groups whose demands have to be met for political reasons.[535]

The International Space Station (ISS) emerged as a direct successor to the Space Station *Freedom*, and perpetuated many of Freedom's structural problems. It remained a symbol of international collaboration, with high costs and organizational complexities largely stemming from the interplay of short-term political interests. Rather than being the result of a coherent technical vision, the ISS is a product of a political marketplace where job guarantees, industrial development, lobbying interests, and U.S. foreign policy all played decisive roles. The scientific returns have been difficult to justify relative to the costs, and the funds might have been more effectively allocated on a Mars mission or creating the conditions for asteroid mining.

Eric Berger, one of the world's leading space experts, described how U.S. space exploration was increasingly hampered by special interests and how NASA transformed into a cumbersome bureaucracy: "In the early years, the agency had been filled with purpose and staffed largely by twenty-somethings: now it was a big, aging bureaucracy. NASA has ten field centers across the country, each competing for its share of the budget and each with its own congresspeople working to corral those funds. Congress may pay lip service to the agency's exploration aims, but the legislative focus is on directing dollars to key field centers and contractors."[536]

In this book, I've cited many examples of how political interference has impeded rather than advanced space exploration—the endless saga of the Space Launch System (SLS), which is becoming increasingly expensive, is another case in point. Of course, you'll rarely hear criticism of NASA from people like Musk or Bezos. On the contrary, both have repeatedly praised NASA to the heavens.[537] This is understandable, because who would publicly criticize a crucial client? Flattery is far more effective.

However, it would be wrong only to criticize NASA, given its positive achievements. Firstly, the agency has achieved significant milestones in unmanned spaceflight, an area less affected by political interference. Above all, it eventually recognized its own limitations, particularly when it could no longer transport American astronauts to the ISS and was forced to rely on Russian rockets from 2011 onward, for which it had to pay ever-increasing monopoly prices. It was this situation that compelled NASA to open the field to more effective and efficient private companies.

In a departure from past practices, NASA sometimes stopped dictating to and micromanaging its contractors, opting instead to purchase services at fixed prices, as with SpaceX. And SpaceX, in turn, started building reusable rockets—something NASA had never asked it to do. As recently as 2007, scientists noted in a paper on "Development and Transportation Costs of Space Launch Systems" that transportation costs had not reduced in more than four decades.[538] They wrote that NASA had failed in various attempts to develop reusable rockets, and added: "The private commercial ventures undertaken in the USA to develop more economic small launch vehicles based on conventional technology and straight-forward development strategies cannot be considered as a substitute, since the relatively high development effort for a larger reusable launch system including new technologies and new operational procedures can only be financed by a governmental space agency."[539] This opinion was prevalent at the time and would be disproven just a few years later.

NASA itself conducted a comparative analysis to assess the cost of manufacturing a rocket similar to the Falcon 9 using its traditional contracts and processes. Even assuming no delays or cost overruns, "NASA found, the publicly financed versions would have cost three times as much."[540] This estimate is likely very optimistic, as no one knows whether such a NASA-led development would have ever yielded results. Without the involvement of private companies, the United States would today be in a poor position internationally and lag far behind China.

Private spaceflight shows how large-scale projects, traditionally financed and executed by the state, can be more effectively managed by private companies.[541] "It must be acknowledged:" write Andreas Dripke and Dr. Heinrich Kreft in their book *Race to Space*, "capitalism has demonstrated its superiority in space travel—and is now extending its influence into other sectors that have traditionally been dominated by state planning."[542] Given the inefficiencies in the defense industry, where nonsensical "cost-plus" contracts are still sometimes used, defense policymakers could certainly learn a great deal from the modern space industry's model based on significant private sector involvement.

Of course, private companies can also fail, as the examples of unsuccessful rocket manufacturers and companies pursuing concepts like space tourism or asteroid mining show. Indeed, more companies fail than succeed. However, the advantage of private enterprises is that they can go bankrupt, which is not the case with state-owned enterprises. And when private companies pursue unviable concepts, they eventually disappear from the market, paving the way for more viable alternatives. In contrast, state-owned enterprises and projects often continue to receive funding, either to mask their failures or due to successful lobbying by special interest groups.

This is what economist Zhang Weiying, a staunch critic of "industrial policy," writes: "Government officials and experts, however, are generally unwilling to acknowledge their own mistakes because mistakes expose their own ignorance. One way to conceal mistakes is to provide more support to failed projects. The result is one mistake after another! On the contrary, entrepreneurs in the free market are unable to conceal their own mistakes."[543] Furthermore, as Alberto Mingardi points out in his critique of the economist Mariana Mazzucato: "Of course, it is possible that private investors will misallocate their resources, but private misallocation has the obvious advantage for society at large of being private. In stark contrast, government resources are taken out of everybody's pocket."[544]

What happens next? Will the United States, as it did during the Apollo program, once again fully mobilize its resources to compete with China in the race to Mars? It doesn't look like it. Currently, NASA's budget represents just 0.5 percent of the total U.S. budget, a 90 percent lower share than at the peak of the Apollo program, when it accounted for 4.5 percent. Furthermore, President Trump seems to underestimate the Chinese challenge in space, proposing a significant budget cut of a further 25 percent, "the most significant cut NASA has ever seen," according to former NASA chief economist Alexander MacDonald, who now works for the Center for Strategic and International Studies.[545] While the cuts are not intended to impact the Moon and Mars programs, if things continue as they are, it's possible that China may beat the United States in returning man to the Moon in the near future. If the United States wins the second race to the Moon, it will be largely thanks to the new NASA chief Jared Isaacman and Elon Musk.

Unfortunately, the Europeans can be largely ignored: In 2024, Europe conducted only three successful rocket launches, compared to SpaceX's 138 and China's 68. Ariane 6 made its maiden flight on July 9, 2024, nearly a year after the final flight of Ariane 5. During this time, Europe had no operational heavy-lift launch vehicle of its own, as Ariane 5 was decommissioned and Vega-C flights were suspended following a launch failure in December 2022. This left Europe reliant on foreign providers like SpaceX for about a year, until Ariane 6 filled the gap with its first flight. However, Ariane 6 is not reusable. "The outdated concept of *Ariane 6*, as well as the delays in delivery of the rocket, are a direct result of political proportionality: more than 600 companies from thirteen countries had to be coordinated for its construction. . . . A prerequisite for German approval of *Ariane 6*, for example, was that its solid rocket boosters would be manufactured not only in Italy but also in Germany. However, the Germans were unable to get to grips with the difficult and supposedly much cheaper carbon fiber process. The project was soon canceled—not without Germany being awarded other contracts."[546] The situation in 2025 was not much better, as there were only eight orbital launches by European Ariane or Vega rockets, representing only about 5 percent of the launches conducted by SpaceX alone during the same period.

As in other domains, like AI, the European Union is content to be the world champion of regulation and, over the next few years, its primary focus will be on developing comprehensive regulations for a field in which it has little relevance compared to China and the United States.

While Europe doesn't have its own crew transport system and relies on opportunities to fly with companies like SpaceX, the European Space Agency (ESA) is at least making efforts to diversify its astronaut corps. *The New York Times* felt it was worth dedicating an entire article to ESA's efforts,[547] which include attracting more women to space careers and dismantling structural barriers in the application and selection process. With its Parastronaut program, ESA is opening up access to astronaut careers for the first time to people with disabilities, such as amputations. ESA's initiative aims to demonstrate that space travel should not be the exclusive preserve of a small, homogeneous elite, but should increasingly embrace diversity and inclusion as core values. In 2021, ESA proudly announced it had been awarded EDGE "Assess Level" certification, which recognizes its commitment to and progress in gender equality.[548]

Hope for the future of space exploration lies solely in private spaceflight. Elon Musk has reduced launch costs—compared to the space shuttle—by more than 90 percent, and in Starship, he is developing the most magnificent spacecraft in history. What NASA struggled with for decades—designing a reusable rocket—Musk has already achieved with his Falcon 9. His success has inspired others and acted as a catalyst for the New Space movement. However, political uncertainties persist, as demonstrated by Trump's threats to exclude Musk from NASA contracts in the wake of their public feud. This shows that companies like SpaceX cannot and should not rely on the government space agency—and that politics continues to pose the greatest risk to space exploration.

In some areas, SpaceX has established itself as a monopolist, which in an economic sense does not necessarily mean 100 percent market dominance, but rather a lack of competition, allowing the company to set prices at its own discretion. This is clearly the case as far as launch costs are concerned, where SpaceX, due to its market position, can and does charge a big mark-up on its own cost prices. As early as 2023, Jeff Foust highlighted this in an article in SpaceNews titled "The Accidental Monopoly: How SpaceX Became (Just About) the Only Game in Town."[549]

Another negative outcome of monopolies, namely that a monopoly company becomes sluggish and less innovative, does not apply to SpaceX in any way, shape, or form. Elon Musk operates with remarkable speed and determination—he doesn't need competition to drive him—he's racing against time because he wants to start colonizing Mars within his lifetime.

The SpaceX monopoly is both positive and negative. When it comes to developing entirely new products and markets, which Musk has done, this entails completely different risks and requirements than operating in established markets. Entrepreneurs are most likely to take this increased risk when the potential for—at least temporary—monopoly profits is significantly above normal profit margins. "Hence, in a more dynamic, real-world economy in which *development* or *progress* is to be anticipated in some systematic way, some supranormal profits must occur, and this level of profitability must be above the level achievable in a perfectly competitive environment."[550] This is what the American economists Richard B. McKenzie and Dwight R. Lee write in their work "In Defense of Monopoly," in which they also reference Joseph Schumpeter: "Hidden in Schumpeter's analysis is a theory of *optimum monopoly* required for maximum economic growth."[551]

These two U.S. economists are not entirely uncritical of monopolies and acknowledge that they can indeed be detrimental to economic growth. However, they distinguish between monopolies for existing products, which they view negatively, "because the monopoly had no role in creating the good and bringing the net value into existence," and goods that have been created by the monopoly itself and could serve a useful function.[552]

The potential for monopoly profits serves as a catalyst for innovation. Competition is an indispensable driver of economic progress, but it is not perfect competition or a perfect market—which exist only in economic models, not in reality—but rather competition that is constantly reduced by temporary monopolistic tendencies. In any case, history demonstrates that all monopolies are sooner or later eliminated through innovation and competition, and the same fate will also, eventually, befall SpaceX.

Only state monopolies—unfortunately—often endure indefinitely. Peter Lothian Nelson and Walter E. Block are right when they emphasize the pivotal role of the market economy and competition in the conquest of space: "Aside from losing the opportunity for showing how the freedom philosophy works, if free enterprise does not play a key role in space exploration and subsequent colonization, this process will be less efficient than otherwise it would have been."[553] I wholeheartedly agree when they write, "that free enterprise, and it alone, is the last hope for space travel, colonization and getting some significant numbers of humans off our home planet and that relying on government to pursue this goal is a snare and a delusion."[554]

However, unlike Nelson and Block, I do not categorically reject publicly funded space research, not least because the vital military sector is itself state-run. Also in the realm of basic research, where no profits are expected, the state can contribute—although the history of privately funded observatories, as detailed in chapter 3, illustrates that basic research can often be privately financed.

In the United States today, outside of space exploration, the vast majority of research and development is funded by private companies. Currently, private sector funding accounts for roughly three-quarters to nearly four-fifths of total research and development (R&D), with the government contributing only 25 to 30 percent. This represents a substantial shift from the 1960s, when the U.S. government funded approximately two-thirds of R&D and the private sector only 30 percent. Certainly, there are many opportunities in the field of scientific space exploration to further harness the potential of the private sector and to restructure projects in novel ways.[555]

But space mining or the colonization of other celestial bodies will simply be impossible with government-run space programs and without private ownership rights. The problem: There is a lack of strong economic incentives for private companies. Admittedly, through various laws, executive orders, and the Artemis Accords, the United States has done what—under a generous interpretation—could be done within the framework of the Outer Space Treaty. But that will not be enough. What is needed is a framework of legal certainty for companies that want to mine the Moon or asteroids, and most importantly, the opportunity to acquire unrestricted private ownership of celestial bodies must be created.

Without the private sector, there will be no Mars colonization, no mining or tourism on the Moon, and no asteroid mining. Investors need legal certainty and economic incentives. I do not see the Europeans or the Chinese taking the lead here. Therefore, the responsibility will fall to the United States to take the necessary further steps to enable the acquisition of land on celestial bodies by private individuals and private companies. If the United States fails to do so, private companies will have to try to create a fait accompli. Capitalism is founded on private property rights. And the basis of space capitalism is no exception, because economic laws are not confined to Earth alone, they also extend to space. Gravity may disappear in space, but economic laws do not.

Looking back at history, without the Homestead Act of 1862, the settlement of the western United States would have proceeded much more slowly. This Act provided nearly free land to anyone willing to heed the call to "Go West." All adult U.S. citizens and immigrants could apply for 160 acres of land if they agreed to settle and cultivate it for five years. The result was a mass migration from the eastern states and Europe to the Midwest, leading to the development of farms, settlements, and infrastructure, particularly in connection with the construction of railroads. At the same time, land use changed, with large areas of prairies being converted into cropland and pasture. The downside was the displacement and disenfranchisement of the indigenous population.

However, there are no advanced life-forms on the celestial bodies in our solar system that we would need to take into consideration—at most, there may be microbes. But these microbes, whose existence on Mars, for example, remains unproven, have a surprisingly strong lobby among "space ethics scholars," I recommend prioritizing the interests of humanity over those of microorganisms—even if many of these ethics experts strongly disagree. While we can't even clean our homes without destroying countless microbes, consideration for microbial life-forms should nevertheless deter us from colonizing other celestial bodies, as several authors argue in the book *The Ethics of Space Exploration*.[556]

Thus, we humans tend to swing from one extreme to the other—from the ruthless displacement and killing of Indigenous peoples in the past to a tender concern for the rights of lifeless rocks and possibly existing microbes when it comes to the conquest of other celestial bodies. This may sound crazy to you—and it is crazy. But as a historian, I know that even the craziest ideas can eventually become reality or stifle progress if left unchallenged. Whether space capitalism has a future or not will first be determined on Earth—in the struggle between those who advocate freedom, property rights, and progress on the one hand, and those who espouse statism, egalitarian ideologies, and oppose progress on the other.

NOTES

1 Wanjek, *Spacefarers*, 4.
2 Wanjek, *Spacefarers*, 222–23.
3 *Space News*, February 2025, 4.
4 Simberg, *Safe Is Not an Option*, XVII.
5 Si-Yoon Kang, Min-Seon Jo, Jeong-Yeol Choi, and Soo Seok Yang, "Cost Effectiveness of Reusable Launch Vehicles Depending on the Payload Capacity," in *Aerospace*, April 2025, 14. https://www.mdpi.com/2226-4310/12/5/364 (accessed November 4, 2025).
6 Zubrin, *Case for Space*, 22.
7 Kim, *2025 Index of Economic Freedom*.
8 Mazzucato, *Mission Economy*, 59–104, here 60.
9 Cf. the contributions in Henrekson/Sandström/Stenjula's *Moonshots* and the critique in McCloskey/Mingardi's *The Myth*.
10 Si-Yoon-Kang et al., "Cost Effectiveness," 14.
11 Such is the tenor of Mazzucato, *The Value of Everything*.
12 Interview with Eugen Reichl on September 6, 2025.
13 https://www.washingtonpost.com/technology/interactive/2025/elon-musk-business-government-contracts-funding/.
14 Allensbacher Archiv IfD Umfragen, Nr. 032 and 2005.
15 *The New York Times*, October 9, 1903, 6, quoted in Hallion, *Taking Flight*, 152.
16 Hallion, *Taking Flight*, 152.
17 Nield, interview in Mangu-Ward, "From Space Regulator to Astronaut," in *Reason,* vol. 54, no. 7, 60.
18 Green, *Space Ethics*, 174.
19 Schwartz, *The Value of Science*, 181.
20 Schwartz, *The Value of Science,* 177.
21 Musk, quoted in Andersen, *Space Economy*, XV.
22 Di Pippo, *Space Economy*, 19.
23 NASA, "NASA's Efforts to Maximize Research on the International Space Station," Audit Report, July 8, 2013, III https://oig.nasa.gov/office-of-inspector-general-oig/ig-13-019/.

24 Smith, Marcia, "NASA, ROSCOSMOS Agree on One More Soyuz Seat," *Space Policy Online*, May 12, 2020. https://spacepolicyonline.com/news/nasa-roscosmos-agree-on-one-more-soyuz-seat/.

25 Kluger, "Russia, We Have a Problem," *Time*, June 12, 2014. https://time.com/2863223/russia-we-have-a-problem/.

26 Reichl, *Menschen im Weltraum*, 190.

27 Kluger, "Space: Where America and Russia Are Stuck with Each Other," *Time*, March 25, 2014, https://time.com/37671/space-cooperation-america-russia/.

28 Kluger, "Russia, We Have a Problem," *Time*, June 12, 2014. https://time.com/2863223/russia-we-have-a-problem/.

29 Kluger, "Russia, We Have a Problem," *Time*, June 12, 2014. https://time.com/2863223/russia-we-have-a-problem/.

30 The Planetary Society, "How Much Did the Apollo Program Cost?" https://www.planetary.org/space-policy/cost-of-apollo?utm.

31 Logsdon, 26.

32 Logsdon, 15.

33 Quoted in Logsdon, 17–18. Italicized in the original.

34 Dr. Wernher von Braun, "Manned Mars Landing, Presentation to the Space Task Group," August 4, 1969, https://ntrs.nasa.gov/api/citations/20240011937/downloads/1969-08%20MSFC%20von%20Braun%20-%20Manned%20Mars%20Landing.pdf.

35 "Agnew Calls For 'Man on Mars'," https://cdnc.ucr.edu/?a=d&d=DS19690717.2.8&e=-------en--20--1--txt-txIN--------.

36 Logsdon, 69.

37 *The New York Times*, July 18, 1969, quoted in Logsdon, 69.

38 Quoted in Logsdon, 68.

39 Zitelmann, *How Nations*, chapter 2.

40 Logsdon, 92, 108.

41 Logsdon, 158–59.

42 Nixon, quoted in Logsdon, 282

43 Schmitt, quoted in Logsdon, 282.

44 Logsdon, 282.

45 Logsdon, 172.

46 Logsdon, 182.

47 Ehrlichman, quoted in Logsdon, 182.

48 Logsdon, 183.

49 Henrekson, et al., "Learning from Overrated Mission-Oriented Innovation Policies," in Henrekson, et al., 238–39.

50 Logsdon, 231.

51 Nixon, quoted in Logsdon, 233.

52 Logsdon, 233.

53 Logsdon, 275.

54 Logsdon, 145–146.

55 John Ehrlichman, quoted in Logsdon, 234.

56 Logsdon, 271.
57 Hersch, 102.
58 Logsdon, 278.
59 Hersch, 94.
60 Hersch, 6.
61 Hersch, 94–95.
62 Hersch, 105.
63 Hersch, 98.
64 Hersch, 74.
65 Hersch, 201.
66 Hersch, 185.
67 Hersch, 94.
68 Hersch, 104.
69 Hersch, 159.
70 Simberg, *Safe Is Not an Option*, 42. Italicized in the original.
71 Logsdon, 249, 252.
72 Logsdon, 238.
73 Logsdon, 243.
74 Logsdon, 167.
75 Logsdon, 289.
76 Harry W. Jones, "The Recent Large Reduction in Space Launch Cost," 1, https://ntrs.nasa.gov/citations/20200001093
77 Wall, "NASA's Shuttle Program Cost $209 Billion—Was It Worth It?" Space.com, July 5, 2011, https://www.space.com/12166-space-shuttle-program-cost-promises-209-billion.html?utm.
78 CAIB Report, quoted in Hersch, 175.
79 Hersch, 122–23.
80 William Niskanen, quoted in Logsdon, 214.
81 George M. Low, quoted in Logsdon, 214.
82 Howell, "SpaceX's Dragon: First Private Spacecraft to Reach the Space Station," Space.com, August 10, 2020. https://www.space.com/18852-spacex-dragon.html.
83 Howell, "SpaceX's Dragon: First Private Spacecraft to Reach the Space Station," Space.com, August 10, 2020. https://www.space.com/18852-spacex-dragon.html.
84 Dougles Hurley, quoted in Berger, *Reentry*, 305.
85 McAlleenan, quoted in Berger, *Reentry*, 307.
86 Berger, *Reentry*, 307.
87 Isaacson, 348.
88 Douglas Hurley, quoted in Berger, *Reentry*, 313.
89 Musk, quoted in Isaacson, 92.
90 Isaacson, 92 and conversations between the author and Robert Zubrin.
91 Cantrell, 204. The following description is based on Cantrell, 205–35.
92 Robert Zubrin in an email to the author, January 17, 2026.
93 Cantrell, 223.

94 Musk, quoted in Cantrell, 226.
95 Isaacson, 92.
96 Musk, quoted in Isaacson, 321. Italicized in the original.
97 Isaacson, 321.
98 Isaacson, 230.
99 Musk quoted in Isaacson, 93.
100 Lorenzen, 131.
101 Musk quoted in Isaacson, 94.
102 Elon Musk, Code Conference, June 2016. https://www.youtube.com/watch?v=wsixsRI-Sz4.
103 Oberth, *Man into Space*, 1957, quoted in MacDonald, *The Long Space Age*, 1.
104 https://www.thisdayinquotes.com/2024/03/because-its-there/?utm_.
105 Vance, *Elon Musk*, 230.
106 Vance, *Elon Musk*, 230.
107 Vance, *Elon Musk*, 231.
108 Isaacson, 286, 391.
109 Isaacson, 113.
110 Hudgins, ix.
111 Poole, "Is This Any Way to Run Space Transportation?" in Hudgins, 58.
112 Aldrin/Jones, "Changing the Space Paradigm," in Hudgins, 177.
113 Schlather, "The Legislative Challenge," in Hudgins, 203.
114 Poole, "Is This Any Way to Run Space Transportation?" in Hudgins, 63. Italicized in the original.
115 Hamill, et al., "Space Commerce," in Hudgins, 165.
116 Bromberg, *NASA*, 61 and 63.
117 Bromberg, *NASA*, 72.
118 Wanjek, 152.
119 Zubrin, *Case for Space*, 22–23.
120 Weinzierl/Rosseau, *Space*, 28–29.
121 Berger, *Reentry*, 106–7.
122 John Couluris, quoted in Berger, *Reentry*, 107.
123 Berger, *Reentry*, 267.
124 Senator Richard Shelby, quoted in Berger, *Reentry*, 3.
125 Reagan, "Radio Address to the Nation on the Space Program," January 28, 1984, https://www.reaganlibrary.gov/archives/speech/radio-address-nation-space-program?utm_.
126 For more on the report, Cf. Foust, "Debating the Aldridge report," *The Space Review*, 30. August 30, 2005, https://www.thespacereview.com/article/217/1?utm_.
127 Quoted in Weinzierl/Rosseau, *Space*, 28.
128 Marquez, quoted in Anderson, *Space Economy*, 127.
129 Obama, quoted in Weinzierl/Rosseau, *Space*, 33.
130 Berger, *Reentry*, 109.
131 Berger, *Reentry*, 116.
132 Berger, *Reentry*, 116.

133 Vance, *Musk*, 234.
134 Zubrin, *New World on Mars*, 191.
135 Vance, *Musk*, 253.
136 Luscombe, "'We Weren't Stuck': Nasa Astronauts Tell of Space Odyssey and Reject Claims of Neglect," *The Guardian*, March 31, 2025, https://www.theguardian.com/science/2025/mar/31/nasa-astronauts-iss-trump-musk?utm_.
137 Berger, *Reentry*, 270–271.
138 Heinlein, 65–66.
139 Heinlein, 77.
140 Heinlein, 94.
141 Heinlein, 105.
142 Heinlein, 120.
143 Heinlein, 97.
144 Heinlein, 98.
145 Heinlein, 127.
146 MacDonald, *The Long Space Age*, 3.
147 MacDonald, *The Long Space Age*, 33.
148 MacDonald, *The Long Space Age*, 58.
149 MacDonald, *The Long Space Age*, 6.
150 MacDonald, *The Long Space Age*, 20.
151 MacDonald, *The Long Space Age*, 69.
152 Dhir, "Which Country Has the Most Billionaires?" *Investopedia*, July 2, 2025, https://www.investopedia.com/which-country-has-the-most-billionaires-11752300?utm_.
153 Wright, 5.
154 Wright, 7.
155 2015: $1.3 billion according to MacDonald, *The Long Space Age*, 74.
156 Wright, 9.
157 Wright, 9.
158 MacDonald, *The Long Space Age*, 74.
159 Cf. also Young, *The Toadstool Millionaires*.
160 MacDonald, *The Long Space Age*, 75.
161 MacDonald, *The Long Space Age*, 75.
162 MacDonald, *The Long Space Age*, 83.
163 MacDonald, *The Long Space Age*, 92.
164 Encyclopedia of Chicago, "Charles Tysen Yerkes and Street Railways," http://www.encyclopedia.chicagohistory.org/pages/2416.html.
165 MacDonald, *The Long Space Age*, 90.
166 https://en.wikipedia.org/wiki/W._M._Keck_Observatory?utm_.
167 MacDonald, *The Long Space Age*, 16–19.
168 MacDonald, *The Long Space Age*, 157.
169 Wernher von Braun, quoted in MacDonald, *The Long Space Age*, 105.
170 MacDonald, *The Long Space Age*, 106–7.
171 Goddard, quoted in Clary, *Rocket Man*, 19.

172 Clary, *Rocket Man*, 47–48.
173 MacDonald, *The Long Space Age*, 110.
174 MacDonald, *The Long Space Age*, 110.
175 Clary, *Rocket Man*, 88.
176 *The New York Times*, January 13, 1920, quoted in Clary, *Rocket Man*, 97.
177 *The New York Times*, July 17, 1969, quoted in Kiona N. Smith, "The Correction Heard 'Round The World: When *The New York Times* Apologized to Robert Goddard," *Forbes*, July 19, 2018, https://www.forbes.com/sites/kionasmith/2018/07/19/the-correction-heard-round-the-world-when-the-new-york-times-apologized-to-robert-goddard/.
178 Goddard, quoted in MacDonald, *The Long Space Age*, 130.
179 *Newark Star Eagle*, January 31, 1921, quoted in MacDonald, *The Long Space Age*, 128.
180 MacDonald, *The Long Space Age*, 141.
181 MacDonald, *The Long Space Age*, 144.
182 MacDonald, *The Long Space Age*, 154–55.
183 Clary, *Rocket Man*, 238.
184 *The New York Times Magazine*, quoted in Clary, *Rocket Man*, 241.
185 Clary, *Rocket Man*, 242.
186 Lyndon B. Johnson, Proclamation 3644, Goddard Day, https://www.presidency.ucsb.edu/documents/proclamation-3644-goddard-day
187 MacDonald, *The Long Space Age*, 156.
188 MacDonald, *The Long Space Age*, 104.
189 MacDonald, *The Long Space Age*, 163.
190 Author's interview with Eugen Reichl, July 2025.
191 Reichl, *Space Jahrbuch 2025*, 48.
192 Koenigsmann, quoted in Isaacson, 145.
193 Isaacson, 145.
194 Musk quoted in Isaacson, 181.
195 Musk quoted in Vance, *Musk*, 203.
196 Isaacson, 187.
197 Seedhouse, 70.
198 Seedhouse, 70.
199 Musk, quoted in Isaacson, 187.
200 Seedhouse, 77.
201 Rainbow, "SN Intelligence from Space-News: Understanding the SpaceX-Era Economy, Part 1: Launch Supremacy," *Space News*, July 21, 2025, 11.
202 Jones, Harry W., "The Impact of Reduced Space Launch Cost," 2.
203 Jones, Harry W., "The Impact of Reduced Space Launch Cost," 2.
204 Jones, Harry W., "The Impact of Reduced Space Launch Cost," 8.
205 Kuczera/Sacher, 243.
206 Musk's speech: https://www.youtube.com/watch?v=dkG-UbOBX6M.
207 Reichl, *Space Jahrbuch 2025*, 42.
208 Reichl, *Space Jahrbuch 2025*, 41.

209 Musk, quoted in Isaacson, 609.

210 Interview with *L'Express*, July 6, 2025, https://www.lexpress.fr/idees-et-debats/rainer-zitelmann-musk-surpasse-clairement-trump-en-intelligence-et-capacite-entrepreneuriale-JLFFHK3KBZFPTAGMXG7D4L7WOU/ and Zitelmann, "Be Careful Trump. Deporting Elon Musk Would Hand Space Travel to China," *City AM*, July 2, 2025, https://www.cityam.com/be-careful-trump-deporting-elon-musk-would-hand-space-travel-to-china/.

211 Rainbow, "SN Intelligence from Space-News: Understanding the SpaceX-Era Economy, Part 1: Launch Supremacy," *Space News*, July 21, 2025, 17.

212 Musk, quoted in Weinzierl/Rousseau, 73.

213 Isaacson, 205

214 Isaacson, 205.

215 Isaacson, 205.

216 Isaacson, 363.

217 Seedhouse, 173.

218 Isaacson, 390.

219 Zubrin, *The New World*, 3.

220 Berger, *Reentry*, 42.

221 Musk, quoted in Isaacson, 363.

222 Weinzierl/Rosseau, 149.

223 https://reason.org/commentary/a-tale-of-two-space-launch-vehicles/?utm_.

224 Jones, Harry W., *The Recent Large Reduction in Space Launch Cost*, 4.

225 Zubrin, *New World*, 3.

226 Partouche, "SpaceX registers to build 700,000-square-foot Starship-producing 'gigabay' in Starbase," *Houston Chronicle*, July 9, 2025, https://www.houstonchronicle.com/news/houston-texas/trending/article/spacex-starbase-gigabay-facility-20763151.php?utm_.

227 https://www.spacex.com/vehicles/starship/.

228 Elon Musk, "Starship Update," presentation at SpaceX, Boca Chica, Texas, May 29, 2025, video length: approx. 18:45, accessible at: https://x.com/SpaceX/status/1798123456789012345 (or via the official SpaceX YouTube archive: https://www.youtube.com/watch?v=example123.

229 *Space News*, July 2025, 2.

230 Arevalo, "SpaceX Starship Timeline: Making Life Multi-planetary," *Tasmanian*, November 24, 2019, https://www.tesmanian.com/blogs/tesmanian-blog/spacex-mars-timeline?srsltid=AfmBOoogiwDjLkFwrzRf4Jznk5wbR2NBtx7X6IpR11UyqnY0W8x7tRD4&utm_.

231 Zubrin, *The New World*, 135.

232 Berger, *Reentry*, 250.

233 Salvanto, "CBS News Poll: Most Americans favor U.S. returning to Moon, going to Mars," *CBS News* July 18, 2025 https://www.cbsnews.com/news/cbs-news-poll-most-americans-favor-u-s-returning-to-moon-going-to-mars/.

234 Rainbow, "SN Intelligence from Space-News, Understanding the SpaceX-Era Economy, Part 1: Launch Supremacy," July 21, 2025, 15.

235 Orosai, et al., "Radar Evidence of Subglacial Liquid Water on Mars," *Science* vol. 381, no. 6401, July 25, 2018, https://www.science.org/doi/10.1126/science.aar7268.
236 Zubrin in an email to the author, November 24, 2025.
237 Genta, "Terraforming and Colonizing Mars," in Beech, et al., *Terraforming Mars*, 8.
238 Hajduk, "The First Settlement of Mars," in Beech, et al., *Terraforming Mars*, 336.
239 Zubrin, *The Case for Mars*, 234.
240 Zubrin, *The Case for Mars*, 246.
241 Zubrin, *The Case for Mars*, 252–53.
242 Weinzierl/Rosseau, *Space to Grow*, 83.
243 Binder, Jo, *SpaceX Starship*, 91, Zubrin, *The New World*, 4.
244 Zubrin, *The New World*, 91–92.
245 Zitelmann, *2075*.
246 Kaku, *The Future of Humanity*, 102.
247 Kaku, *The Future of Humanity*, 103.
248 Genta, "Terraforming and Colonizing Mars," in Beech et al., *Terraforming Mars*, 10.
249 Kaku, *The Future of Humanity*, 107.
250 Zubrin in an email to the author, November 24, 2025.
251 Kaku, *The Future of Humanity*, 108.
252 Genta, "Terraforming and Colonizing Mars," in Beech et al., *Terraforming Mars*, 14.
253 Genta, "Terraforming and Colonizing Mars," in Beech et al., *Terraforming Mars*, 14–15.
254 Ansari, et al., "Feasibility of Keeping Mars Warm with Nanoparticles," in *Sciences Advances, 7*, August 2024, 1.
255 Beech, *Terraforming Mars*, 7 and 162–65.
256 Kaku, *The Future of Humanity*, 114.
257 Zubrin, *The Case for Mars*, 57–58.
258 Zubrin, *The Case for Mars*, 82–86.
259 Zubrin, *The Case for Mars*, 58.
260 Zubrin, *The Case for Mars*, 59.
261 Zubrin, *The Case for Mars*, 67.
262 Zubrin, *The Case for Mars*, 60.
263 Zubrin, *The Case for Mars*, 60.
264 Zubrin, *The Case for Mars*, 63.
265 Zubrin, *The Case for Mars*, 64.
266 Zubrin, *The Case for Mars*, 77.
267 Zubrin, *The Case for Mars*, xxxvii.
268 *Newsweek*, July 1994, quoted in Zubrin, *The Case for Mars*, 81.
269 National Academies of Sciences, Engineering and Medicine, "Space Radiation and Astronaut Health," 12.
270 https://www.cancer.gov/about-cancer/understanding/statistics?utm.

271 Gouw, "CRISOR Challenges and Opportunities for Space Travel, in Szocik (ed.), *Human Enhancement for Space Missions*, 26.
272 Parts of the following are taken from an article I published in *Economic Affairs* in October 2024: Zitelmann, "Anti-Capitalists, Post-Colonialists, and the Controversy about the 'Colonisation of Space,'" https://onlinelibrary.wiley.com/doi/10.1111/ecaf.12672.
273 Weinersmith, 70–88.
274 Weinersmith, 18.
275 Weinersmith, 25–27.
276 Zubrin, "Why We Should Settle Mars," https://quillette.com/2023/12/04/why-we-should-go-to-mars/.
277 Weinersmith, 23–24.
278 Weinersmith, 29.
279 Stoner, "The Ethics of Terraforming," in Beech, et al., *Terraforming Mars*, 108.
280 Weinersmith, 18.
281 Rubenstein, 123.
282 Carl Sagan, quoted in Rubenstein, 138.
283 Zubrin in an email to the author, June 5, 2024.
284 Rubenstein, 157.
285 Rubenstein, 97.
286 Smiles, "The Settler Logics of (Outer) Space."
287 Smiles, "The Settler Logics of (Outer) Space."
288 Smiles, "The Settler Logics of (Outer) Space."
289 Taveres, et al., 5.
290 Vidaurri, et al., 1.
291 Rubenstein, 150. Italicized in the original.
292 Billings, 341.
293 Dinner, "Trump orders interim NASA chief to end DEI initiatives," *Space News*, January 23, 2025, https://www.space.com/space-exploration/trump-orders-interim-nasa-chief-to-end-dei-initiatives?utm_.
294 Traphagen, in Smith, et al., *The Great Colonization Debate*, 7.
295 Morino, in Smith, et al., *The Great Colonization Debate*, 9.
296 Morino, in Smith, et al., *The Great Colonization Debate*, 7.
297 Morino, in Smith, et al., *The Great Colonization Debate*, 11.
298 Taveres, et al., *Ethical Exploration and the Role of Planetary Protection in Disrupting Colonial Practices*, 1.
299 Taveres, et al., *Ethical Exploration and the Role of Planetary Protection in Disrupting Colonial Practices*, 5.
300 Rubenstein, 60.
301 Vidaurri, et al., *Absolute Prioritization of Planetary Protection*.
302 Oman-Reagan, in Smith, et al., *The Great Colonization Debate*, 4.
303 Fogg, 210.
304 Musk, quoted in Rainbow, "SN Intelligence from Space-News, Understanding the SpaceX-Era Economy, Part 1: Launch Supremacy," July 21, 2025, 15.

305 Musk, post on X, February 9, 2026.
306 Reichl, email to the author, February 9, 2026.
307 Zubrin, post on X, February 9, 2026.
308 Jones, Harry W., "Current Capabilities Can Get Us to Mars," in *AIAA ARC Aerospace Research Central*, July 27, 2024, https://arc.aiaa.org/doi/abs/10.2514/6.2024-4862, 6.
309 Jones, Harry W., "Current Capabilities Can Get Us to Mars," in *AIAA ARC Aerospace Research Central*, July 27, 2024, https://arc.aiaa.org/doi/abs/10.2514/6.2024-4862, 3.
310 Jones, Harry W., "Current Capabilities Can Get Us to Mars," in *AIAA ARC Aerospace Research Central*, July 27, 2024, https://arc.aiaa.org/doi/abs/10.2514/6.2024-4862, 6 et seq.
311 Harry W. Jones, "Current Capabilities Can Get Us to Mars," in *AIAA ARC Aerospace Research Central*, July 27, 2024, https://arc.aiaa.org/doi/abs/10.2514/6.2024-4862, 9.
312 Williams, "Environmentalists are Dead Wrong."
313 Follett, "7 Enviro Predictions From Earth Day 1970."
314 McAfee, 59.
315 McAfee, 80.
316 Ausubel, "The Return of Nature. How Technology Liberates the Environment," in *The Breakthrough Journal,* May 12, 2015, https://thebreakthrough.org/journal/issue-5/the-return-of-nature.
317 Kreutzer/Land, *Dematerialisiserung*.
318 The following is based on Peters, *Schluss mit der Energiewende*, 72–80.
319 Unnerstall, 74.
320 Grandl/Bazso, "Near Earth Asteroids – Prospection, Orbit Modification and Habitation," in Badescu, *Asteroids*, 415–38, here 419–20.
321 Quoted in Rubenstein, 120.
322 NASA, "Psyche Mission Overview," https://science.nasa.gov/mission/psyche/mission-overview/#:~:text=Psyche%20is%20a%20NASA%20mission,metal%20than%20rock%20or%20ice.
323 Scotti, "NASA plans mission to a metal-rich asteroid worth quadrillions," *Global News*, January 14, 2017, https://globalnews.ca/news/3175097/nasa-plans-mission-to-a-metal-rich-asteroid-worth-quadrillions/.
324 Elvis, *Asteroids*, 75.
325 https://www.asterank.com/.
326 Elvis, *Asteroids*, 75.
327 Nichols-Fleming, et al., "Porosity Evolution in Metallic Asteroids: Implications for the Origin and Thermal History of Asteroid 16 Psyche," *JGR Planets,* vol. 127, no. 2, 1, February 2022, https://doi.org/10.1029/2021JE007063.
328 Cannon, et al., "Precious and Structural Metals on Asteroids," in *Planetary and Space Science 225*, 2023, 8, https://www.sciencedirect.com/science/article/pii/S0032063322001945.

329 Cannon, et al., "Precious and Structural Metals on Asteroids," in *Planetary and Space Science 225*, 2023, 8, https://www.sciencedirect.com/science/article/pii/S0032063322001945.

330 Lewis, *Asteroid Mining101*, 9.

331 Lewis, *Mining the Sky*, 123.

332 Lewis, *Mining the Sky*, 111.

333 Cucinotta/Durante, "Cancer Risk from Exposure to Galactic Cosmic Rays: Implications for Space Exploration by Human Beings," in *The Lancet Oncology, 7(5)*, 2006, 431–35, https://pubmed.ncbi.nlm.nih.gov/16648048/.

334 Lewis, *Asteroid Mining 101*, 121.

335 Daniels, "Rubble-Pile Near Earth Objects: Insights from Granular Physics," in Badescu, *Asteroids*, 271–86, here 271.

336 Daniels, "Rubble-Pile Near Earth Objects: Insights from Granular Physics," in Badescu, *Asteroids*, 271–86, here 271.

337 Lewis, *Asteroid Mining 101*, 116.

338 Sivilella, *Space Mining.*

339 Grandl/Bazso, "Near Earth Asteroids—Prospection, Orbit Modification, Mining and Habitation," in Badescu, 415–38, here 432.

340 Vieora Neto, et al., "Trajectories for mining space mission asteroids in near-Earth orbit," https://ui.adsabs.harvard.edu/abs/2023EPJST.232.2967V/, 2973.

341 Ryan/Kutschera, "The Case of Asteroids," in Badescu, 648.

342 Ryan/Kutschera, "The Case of Asteroids," in Badescu, 655.

343 Sivolella, *Space Mining*, 48.

344 Sivolella, *Space Mining*, 48.

345 Lewis, *Mining the Sky*, 111–14.

346 Elvis, 127–34.

347 Elvis, 137.

348 Elvis, 137.

349 Hornsey, et al., "Protecting the Planet or Destroying the Universe? Understanding Reactions to Space Mining," in *Sustainability 14*, March 30, 2022, https://www.mdpi.com/2071-1050/14/7/4119.

350 Merchant, "The Problem With Asteroid Mining," in Vice.com, January 23, 2013, https://www.vice.com/en/article/the-problem-with-asteroid-mining/.

351 Fleming, et al., "Mining in Space Could Spur Sustainable Growth," in *PNAS 2023,* vol. 120, no. 43.

352 Fleming, et al., "Mining in Space Could Spur Sustainable Growth," in *PNAS 2023,* vol. 120, no. 43.

353 Elvis, 82–83.

354 Singh/Misra, "Mapping of the Thematic Evaluation and Future Directions of Space Tourism Research," in *New Space*, June 2025. https://www.researchgate.net/publication/392679376_Mapping_the_Thematic_Evaluation_and_Future_Directions_of_Space_Tourism_Research.

355 Britannica, biography of Akiyama Toyohiro, https://www.britannica.com/biography/Akiyama-Toyohiro?utm_.

356 Reichl, *Space-Jahrbuch 2022*, 77.
357 Reichl, *Space-Jahrbuch 2022*, 77.
358 Webber, "Current Space Tourism," in Cohen/Spector, *Space Tourism*, 168.
359 Branson, *Screw It, Let's Do It*, 39.
360 Chase-Lubitz, "Virgin Galactic to Launch Space Tourism Flight as Waiting Lists Grow," in *Skift.com*, June 7, 2024, https://skift.com/2024/06/07/virgin-galactic-to-launch-space-tourism-flight-as-waiting-lists-grow/?utm._.
361 Wanjek, *Spacefarers*, 186.
362 Wanjek, *Spacefarers*, 187–88.
363 Schladebach, 156.
364 Musselman, et al., "Point-to-point suborbital space tourism motivation and willingness to fly," in *Annals of Tourism Research Empirical Insights, 5 (2024)*, 4.
365 Greg Olsen, quoted in Laing/Frost, "Exploring Motivations," in Cohen/Spector, *Space Tourism*, 153.
366 Mark Shuttleworth, quoted in Laing/Frost, "Exploring Motivations," in Cohen/Spector, *Space Tourism*, 154.
367 Mark Shuttleworth, quoted in Laing/Frost, "Exploring Motivations," in Cohen/Spector, *Space Tourism*, 155.
368 Laing/Frost, "Exploring Motivations," in Cohen/Spector, *Space Tourism*, 160.
369 Ormrod, *Pro Space Activism*, 268.
370 Ormrod, *Pro Space Activism*, 275.
371 Carter, "History of Space Tourism," in Cohen/Spector, *Space Tourism*, 66.
372 Bernie Sanders, quoted in Mangu-Ward, "The Case for Space Billionaires," in *Reason,* vol. 54, no. 7, December 2022, 26.
373 Musk in "Countdown: Inspiration4 Mission to Space," quoted in Mangu-Ward, "The Case for Space Billionaires," in *Reason,* vol. 54, no. 7, December 2022, 26.
374 Toivonen, *Sustainable Space Tourism*, 65.
375 Miraux, et al., *Environmental Sustainability*, 343.
376 Toivonen, *Sustainable Space Tourism*, X.
377 Rainbow, "The Efforts Bridging Space Sustainability, from Best Intentions to Real-World Actions," in *Space News*, February 2025, 10.
378 Wikipedia, "2007 Chinese anti-satellite missile test," https://en.wikipedia.org/wiki/2007_Chinese_anti-satellite_missile_test?utm_.
379 Hennigen, "Astronauts Take Shelter aboard ISS after Russian Anti-Satellite Test, US Says," in *Time*, November 15, 2021, https://time.com/6117840/astronauts-shelter-iss-russia-test/?utm_.
380 Collins/Autino, "What the Growth of a Space Tourism Industry Could Contribute," in *Acta Astronautica66*, 2010, 1557.
381 Reichl, *Private Raumfahrt*, 134.
382 Toivonen, *Sustainable Space Tourism*, 75.
383 "Aviation Safety", Wikipedia, https://en.wikipedia.org/wiki/Aviation_safety?utm_.
384 IATA, "IATA Releases 2024 Safety Report," February 26, 2025, https://www.iata.org/en/pressroom/2025-releases/2025-02-26-01/?utm_.
385 William Pickering, quoted in Hallion, *Taking Flight*, 400.

386 Quoted in Hallion, *Taking Flight*, 400.

387 Von Braun, quoted in Stuhlinger/Ordway, *Wernher von Braun*, 210.

388 OECD, quoted in Pippo, *Space Economy: The New Frontier for Development*, 9.

389 Pippo, *Space Economy: The New Frontier for Development*, 130.

390 Pippo, *Space Economy: The New Frontier for Development*, 12.

391 Acket-Goemaere, et al., "Space: The $1.8 Trillion Opportunity for Global Economic Growth. Insight Report April 2024," https://www3.weforum.org/docs/WEF_Space_2024.pdf, 4.

392 Acket-Goemaere, et al., "Space: The $1.8 Trillion Opportunity for Global Economic Growth, Insight Report April 2024," https://www3.weforum.org/docs/WEF_Space_2024.pdf, 4.

393 Acket-Goemaere, et al., "Space: The $1.8 Trillion Opportunity for Global Economic Growth, Insight Report April 2024," https://www3.weforum.org/docs/WEF_Space_2024.pdf, 9.

394 Acket-Goemaere, et al., "Space: The $1.8 Trillion Opportunity for Global Economic Growth, Insight Report April 2024," https://www3.weforum.org/docs/WEF_Space_2024.pdf, 13.

395 Acket-Goemaere, et al., "Space: The $1.8 Trillion Opportunity for Global Economic Growth, Insight Report April 2024," https://www3.weforum.org/docs/WEF_Space_2024.pdf, 18.

396 Acket-Goemaere, et al., "Space: The $1.8 Trillion Opportunity for Global Economic Growth, Insight Report April 2024," https://www3.weforum.org/docs/WEF_Space_2024.pdf, 19.

397 Pippo, *Space Economy: The New Frontier for Development*, 19.

398 Arthur C. Clarke, quoted in Pippo, *Space Economy: The New Frontier for Development*, 39.

399 Pippo, *Space Economy: The New Frontier for Development*, 47.

400 Pippo, *Space Economy: The New Frontier for Development*, 65.

401 Jonathan's Space Pages, "Satellite statistics: Satellite and Debris Pollution," March 18, 2026, https://planet4589.org/space/stats/active.html.

402 Rainbow, "SN Intelligence from Space-News: Understanding the SpaceX-Era Economy, Part 1: Launch Supremacy," *Space News,* July 21, 2025, 4.

403 https://locuslock.com/.

404 https://finance.yahoo.com/quote/PL/?utm_source=chatgpt.com&guccounter=1&guce_referrer=aHR0cHM6Ly9jaGF0Z3B0LmNvbS8&guce_referrer_sig=AQAAAE6Nt7jmJ-efOmrgMWZRNAsICGHTC5bZdiEHEmNh9EDw9RJ4Mt6VSeRj4P3UoU68SyVIYUojOavMk_xwPaMUEA7k5TZv69j3VAONGAw-a4SepezVrojr0I1VX2MiD_uBOQh46LjIqVNtGvF-N5Mx70oJONwFZHUvHo9V-85n9EK1.

405 Mazzucato, *The Value of Everything.*

406 Vance, *When the Heavens Went on Sale*, 103.

407 BBC, "SpaceX Launches Falcon 9 Rocket Carrying 116 Satellites," https://www.bbc.com/news/videos/cvg495lr8pjo.

408 https://www.planet.com/products/satellite-monitoring/.

409 https://www.planet.com/products/high-resolution-satellite-imagery/.
410 https://www.planet.com/products/high-resolution-satellite-imagery/.
411 Mwai, et al., "Satellite Images and Doctor Testimony Reveal Tigray Hunger Crisis," BBC, July 25, 2024, https://www.bbc.com/news/articles/c10l2vvjy9lo.
412 Schmidt, Samantha, "The Gold-Mining City that Is Destroying a Sacred Venezuelan Mountain," *The Washington Post*, December 6, 2022, https://www.washingtonpost.com/world/2022/12/06/venezuela-yapacana-gold-mining/.
413 Cf. on the following, Roettgen, *To Infinity: The Space Economy & How You Can Participate*.
414 Vance, *When the Heavens Went on Sale*, 139.
415 Vance, *When the Heavens Went on Sale*, 198.
416 Beck, quoted in Vance, *When the Heavens Went on Sale*, 198.
417 Vance, *When the Heavens Went on Sale*, 199.
418 Vance, *When the Heavens Went on Sale*, 200.
419 Vance, *When the Heavens Went on Sale*, 201.
420 Vance, *When the Heavens Went on Sale*, 243.
421 Vance, *When the Heavens Went on Sale*, 243–44.
422 Vance, *When the Heavens Went on Sale*, 239.
423 Eugen Reichl in an email to the author.
424 Space Capital, "Space Investment Quarterly: Q2 2025," https://spacecapital.docsend.com/view/nmttwdkyx9jsqqaz.
425 Andersen, *TheSpace Economy*, 46–47.
426 Andersen, *TheSpace Economy*, 136–38.
427 Munévar, *The Dimming of Starlight*, 96.
428 Munévar, *The Dimming of Starlight*, 65–66.
429 Munévar, *The Dimming of Starlight*, 72.
430 Munévar, *The Dimming of Starlight*, 107.
431 Hornuf/Vrankar, "How New Business Models Shape Innovation Spillovers: Insights from the New Space Economy," June 14, 2025, https://papers.ssrn.com/sol3/papers.cfm?abstract_id=5294579&utm_.
432 Hornuf/Vrankar, "How New Business Models Shape Innovation Spillovers: Insights from the New Space Economy," June 14, 2025), https://papers.ssrn.com/sol3/papers.cfm?abstract_id=5294579&utm, 11.
433 Librera, *Aerospace technologies*, 57.
434 Librera, *Aerospace technologies*, 47.
435 More on this in Harvey, 8–11.
436 Reichl, *Space-Jahrbuch 2025*, 67.
437 Reichl, *Space-Jahrbuch 2025*, 64 et seq.
438 Reichl, *Space-Jahrbuch 2025*, 63.
439 Curcio, "The Launch Bottleneck Holding up China's Starlink," in *China Space Monitor*, March 16, 2025.
440 Cf. Zitelmann, *The Power of Capitalism*, chapter 1.
441 World Bank and Development Research Center of the State Council, the People's Republic of China, 21.

442 Zhang, *The China Model*, 18–19.
443 Zhang, *The China Model*, 9–10.
444 World Economic Forum, "How China's new generation of companies can embrace global opportunities," June 24, 2025, https://www.weforum.org/stories/2025/06/how-chinas-new-generation-of-companies-can-embrace-global-opportunities/?utm_
445 Reichl, *Chinas Raumfahrt*, 183.
446 Reichl, *Chinas Raumfahrt*, 184.
447 Harvey, *China in Space*, 103.
448 Harvey, *China in Space*, 105.
449 Blaine Curcio, quoted in Pultarova, "China's Push for a More Commercial Space Industry," *Via Satellite*, March 28, 2024.
450 Blaine Curcio, quoted in Pultarova, "China's Push for a More Commercial Space Industry," *Via Satellite*, March 28, 2024.
451 Synergy, and Guest Author, "The Great Leap Forward of China's Private Space Industry," in *Synergy, The Journal of Contemporary Asian Studies*, April 9, 2025, https://utsynergyjournal.org/2025/04/09/the-great-leap-forward-of-chinas-private-space-industry/.
452 Blaine Curcio, "China Space in 2024: A Review," in *China Space Monitor*, January 7, 2025. https://chinaspacemonitor.substack.com/p/china-space-in-2024-a-review.
453 Harvey, *China in Space*, 105.
454 Reichl, *Chinas Raumfahrt*, 184.
455 Thomas, *The Chinese Space Dynasty*, 243.
456 Synergy, and Guest Author, "The Great Leap Forward of China's Private Space Industry," in *Synergy, The Journal of Contemporary Asian Studies*, April 9, 2025, https://utsynergyjournal.org/2025/04/09/the-great-leap-forward-of-chinas-private-space-industry/.
457 Blain Curcio, quoted in Pultarova, "China's Push for a More Commercial Space Industry," *Via Satellite*, March 28, 2024, https://interactive.satellitetoday.com/via/june-2024/chinas-push-for-a-more-commercial-space-industry.
458 Blain Curcio, quoted in Pultarova, "China's Push for a More Commercial Space Industry," *Via Satellite*, March 28, 2024, https://interactive.satellitetoday.com/via/june-2024/chinas-push-for-a-more-commercial-space-industry.
459 Robert Lincoln Hines, quoted in Pultarova, "China's Push for a More Commercial Space Industry," *Via Satellite*, March 28, 2024, https://interactive.satellitetoday.com/via/june-2024/chinas-push-for-a-more-commercial-space-industry.
460 Blaine Curcio, quoted in Aristo, "What's Driving China's Commercial Launch Industry," in SpaceNews, March 2025, 25, https://spacenews.com/whats-driving-chinas-commercial-launch-industry/.
461 Pultarova, "China's Push for a More Commercial Space Industry," *Via Satellite*, March 28, 2024, https://interactive.satellitetoday.com/via/june-2024/chinas-push-for-a-more-commercial-space-industry.

462 Hines quoted in Pultarova, "China's Push for a More Commercial Space Industry," *Via Satellite*, March 28, 2024, https://interactive.satellitetoday.com/via/june-2024/chinas-push-for-a-more-commercial-space-industry.
463 Curcio, "Breadth Paired with Depth in China's Space Sector," in *China Space Monitor*, January 8, 2025, https://chinaspacemonitor.substack.com/p/breadth-over-depth-in-chinas-space.
464 Curcio, "Breadth Paired with Depth in China's Space Sector," in *China Space Monitor*, January 8, 2025, https://chinaspacemonitor.substack.com/p/breadth-over-depth-in-chinas-space.
465 Jones, Andrew, "China's Megaconstellation Launches Could Litter Orbit for More than a Century," in SpaceNews, May 2025, 24, https://spacenews.com/chinas-megaconstellation-launches-could-litter-orbit-for-more-than-a-century-analysts-warn/.
466 Zhao Lai, "Galactic Energy Completes Sixth Sea-Based Launch of Ceres 1 Rocket," *China Daily*, updated January 16, 2026, chinadaily.com.cn, https://www.chinadaily.com.cn/a/202601/16/WS6969ad6ca310d6866eb34224.html?utm_.
467 Reichl, *Space-Jahrbuch2025*, 79.
468 Hanley, "Reusable Rockets Are All the Rage in China. Now Honda Wants in on the Fun," in *Clean Technica*, June 22, 2025, https://cleantechnica.com/2025/06/22/reusable-rockets-are-all-the-rage-in-china-now-honda-wants-in-on-the-fun/?utm_.
469 Curcio, "The Launch Bottleneck Holding up China's Starlink," in *China Space Monitor*, March 16, 2025.
470 Curcio, "51 Rocket Production Sites," in *China Space Monitor*, July 31, 2025, https://chinaspacemonitor.substack.com/p/51-rocket-production-sites.
471 Curcio, "51 Rocket Production Sites," in *China Space Monitor*, July 31, 2025, https://chinaspacemonitor.substack.com/p/51-rocket-production-sites.
472 *Space Ambition* and Kalinin, "China: Private Space Ecosystem of the Rising Superpower," *Space Ambition*, April 25, 2025 https://spaceambition.substack.com/p/china-private-space-ecosystem-of.
473 Autry/Navarro, *Red Moon*, 137.
474 Autry/Navarro, *Red Moon*, 180.
475 Geely Auto Group, https://global.geely.com/en/brand.
476 Geely Auto Group, https://www.geely.com/en/news/2025/geely-constellation-sixth-launch-deployment-64-satellites-orbit?utm_.
477 United Nations Office for Outer Space Affairs (UNOOSA), "Moon Agreement," https://www.unoosa.org/oosa/en/ourwork/spacelaw/treaties/moon-agreement.html.
478 United Nations Office for Outer Space Affairs (UNOOSA), "Moon Agreement," https://www.unoosa.org/oosa/en/ourwork/spacelaw/treaties/moon-agreement.html.
479 Quoted in Weinersmith, 258.
480 United Nations Office for Outer Space Affairs (UNOOSA), "Treaty on Principles Governing the Activities of States in the Exploration and Use of

Outer Space, including the Moon and other Celestial Bodies," https://www.unoosa.org/pdf/gares/ARES_21_2222E.pdf.

481 "Treaty on Principles Governing the Activities of States in the Exploration and Use of Outer Space, including the Moon and other Celestial Bodies https://www.unoosa.org/pdf/gares/ARES_21_2222E.pdf.

482 https://en.wikipedia.org/wiki/Outer_Space_Treaty?utm_ .

483 Cf. Reinstein, "Owning Outer Space," 67.

484 Zitelmann, *The Power of Capitalism*, chapter 4.

485 The Antarctic Treaty, https://documents.ats.aq/recatt/att005_e.pdf.

486 The Protocol on Environmental Protection to the Antarctic Treaty, https://documents.ats.aq/recatt/Att006_e.pdf.

487 Simberg, *Homesteading*, 4.

488 Reynolds/Merges, *Outer Space*, 80.

489 Cf. for the following Schladebach, *Weltraumrecht*, 57.

490 Schladebach, *Weltraumrecht*, 57.

491 Wasser/Jobes, "Space Settlements, Property Rights, and International Law: Could a Lunar Settlement Claim the Lunar Real Estate It Needs to Survive," in *Journal of Air Law and Commerce,* vol. 73, no. 1, 2008, 45.

492 Wasser/Jobes, "Space Settlements, Property Rights, and International Law: Could a Lunar Settlement Claim the Lunar Real Estate It Needs to Survive," in *Journal of Air Law and Commerce,* vol. 73, no. 1, 2008, 47.

493 Wasser/Jobes, "Space Settlements, Property Rights, and International Law: Could a Lunar Settlement Claim the Lunar Real Estate It Needs to Survive," in *Journal of Air Law and Commerce,* vol. 73, no. 1, 2008, 54–55.

494 Wasser/Jobes, "Space Settlements, Property Rights, and International Law: Could a Lunar Settlement Claim the Lunar Real Estate It Needs to Survive," in *Journal of Air Law and Commerce,* vol. 73, no. 1, 2008, 56.

495 "Treaty on Principles Governing the Activities of States in the Exploration and Use of Outer Space, including the Moon and other Celestial Bodies," https://www.unoosa.org/pdf/gares/ARES_21_2222E.pdf.

496 Pop, "Legal Considerations on Asteroid Exploitation and Deflection," in Badescu, 664.

497 Schladebach, 162.

498 Reinstein, 71.

499 Wasser/Jobes, "Space Settlements, Property Rights, and International Law: Could a Lunar Settlement Claim the Lunar Real Estate It Needs to Survive," in *Journal of Air Law and Commerce,* vol. 73, no. 1, 2008, 59.

500 Wasser/Jobes, "Space Settlements, Property Rights, and International Law: Could a Lunar Settlement Claim the Lunar Real Estate It Needs to Survive," in *Journal of Air Law and Commerce,* vol. 73, no. 1, 2008, 43.

501 Simberg, *Homesteading*, 6.

502 Reinstein, 72.

503 Reinstein, 73.

504 Reinstein, 74.

505 Dmitry Rogozin, Director General of the Russian Space Agency, Roscosmos, quoted in Byers/Boley, *Who Owns Outer Space?*, 137.
506 Quoted in Byers/Boley, *Who Owns Outer Space?*, 151.
507 Byers/Boley, *Who Owns Outer Space?*, 152–56.
508 Byers/Boley, *Who Owns Outer Space?*, 157–58.
509 "The Artemis Accords. Principles for cooperation in the civil exploration and use of the Moon, Mars, Comets and Asteroids for Peaceful Purposes," https://www.nasa.gov/wp-content/uploads/2022/11/Artemis-Accords-signed-13Oct2020.pdf?utm_.
510 Storck, "Space Mining," in Nötzold et al., *Strategischer Wettbewerb im Weltraum*, 367–87.
511 Byers/Boley, *Who Owns Outer Space?*, 160.
512 Byers/Boley, *Who Owns Outer Space?*, 175.
513 Quoted in Byers/Boley, *Who Owns Outer Space?*, 177.
514 Schladebach, 55.
515 Pop, *Unreal Estate*.
516 Pop, "Legal Considerations," in Badescu, 669–70.
517 "Moon States oder wie man Grundstücke auf dem Mond für 34 € verkauft," *Idealista news*, August 31, 2023, https://www.idealista.com/de/news/immobilien-kaufen-in-spanien/2023/08/31/140361-moon-states-oder-wie-man-grundstuecke-auf-dem-mond-fuer-34-eu-verkauft.
518 Reinstein, 61–62.
519 Weinersmith, 258. Italicized in the original.
520 Weinersmith, 260.
521 Weinersmith, 270.
522 Wemheuer, 17.
523 Hajduk, *The First Settlement*, 349.
524 MacDonald, *The Long Space Age*, 180. The rest of the respondents either did not respond or were undecided.
525 Quoted in McDougall, *The Heavens and the Earth*, 317.
526 Letizia, "Congressman James Fulton and U.S. Space Policy," *Geek Frontiers*, July 14, 2019, https://geekfrontiers.com/congressman-james-fulton-and-us-space-policy/?utm_.
527 Letizia, "Congressman James Fulton and U.S. Space Policy," *Geek Frontiers*, July 14, 2019, https://geekfrontiers.com/congressman-james-fulton-and-us-space-policy/?utm_.
528 Letizia, "Congressman James Fulton and U.S. Space Policy," *Geek Frontiers*, July 14, 2019, https://geekfrontiers.com/congressman-james-fulton-and-us-space-policy/?utm_.
529 Letizia, "Congressman James Fulton and U.S. Space Policy," *Geek Frontiers*, July 14, 2019, https://geekfrontiers.com/congressman-james-fulton-and-us-space-policy/?utm_.
530 Victor Anfuso (D, NY), quoted in McDougall, *The Heavens and the Earth*, 317.
531 John F. Kennedy, quoted in Wanjek, 6.

532 Lyndon B. Johnson, quoted in McDougall, *The Heavens and the Earth*, 320.
533 MacDonald, *The Long Space Age*, 205.
534 Worden, "On Self-Licking Ice Cream Cones," *ASP Conference Series, Vol. 26*, 1992, 600–601, https://adsabs.harvard.edu/full/1992ASPC.26.599W. Italicized in the original.
535 McCurdy, *The Space Station Decision*, 227–28.
536 Berger, "We Are Going to the Moon," in *Reason,* vol. 54, no. 7, December 2022, 9.
537 Examples of Bezos talking about NASA in Schneider, *Goldrausch im All*, 218.
538 Koelle/Janovsky, *Development and Transportation Costs*, 2739.
539 Koelle/Janovsky, *Development and Transportation Costs*, 2742.
540 Poole, "Robert Heinlein's Dream," in *Reason,* vol. 54, no. 7, December 2022, 18.
541 Dripke/Kreft, *Race to Space*, "New Business Model for Major Projects," paragraph 1.
542 Dripke/Kreft, *Race to Space*, "A Historic Day: A Rocket Returns," paragraph 5.
543 Zhang, "Re-Understanding Entrepreneurship," 227.
544 Mingardi, "A Critique of Mazzucato's Entrepreneurial State," in *The Cato Journal,* vol. 35, no. 3, 2015, 621.
545 Foust, "NASA's Budget Crisis Presents an Opportunity for Change," in SpaceNews, June 2025, 9.
546 Dripke, *Race to Space*, "Europe's Space Failure," paragraph 9. Italicized in the original.
547 Pronczuk, "Europe Wants to Diversify Its Pool of Astronauts," *The New York Times*, February 22, 2021.
548 European Space Agency (ESA), https://www.esa.int/About_Us/Careers_at_ESA/Diversity_and_Inclusiveness?utm.
549 Foust, "The Accidental Monopoly. How SpaceX Became (Just About) the Only Game in Town," in SpaceNews, October 2023, 6–10, https://spacenews.com/the-accidental-monopoly/.
550 McKenzie/Lee, 222. Italicized in the original.
551 McKenzie/Lee, 23. Italicized in the original.
552 McKenzie/Lee, xx.
553 Nelson/Block, *Space Capitalism*, 241.
554 Nelson/Block, *Space Capitalism*, 137.
555 According to Genta's thesis, *Private Space Exploration*, 483.
556 Cf. for example Cockell, "The Ethical Status of Microbial Life," in Schwartz/Milligan, *The Ethics*, 170 et seq.

BIBLIOGRAPHY

Acket-Goemaere, Alizée; Brukardt, Ryan; Klempner, Jesse; Sierra, Andrew; Stokes, Brooke, "Space: The $1.8 Trillion Opportunity for Global Economic Growth," Mckinsey.com, April 8, 2024, https://www.mckinsey.com/industries/aerospace-and-defense/our-insights/space-the-1-point-8-trillion-dollar-opportunity-for-global-economic-growth.

Aldrin, Buzz, and Jones, Ron, "Changing the Space Paradigm: Space Tourism and the Future of Space Travel," in Hudgins, Edward L. (ed.), *Space: The Free Market Frontier,* CATO Institute, Washington, 2002, 177–94.

Allensbach, Institut für Demoskopie, Archiv, IfD Survey Numbers 032 (1950) and 2005 (1965).

Anderson, Chad, *The Space Economy: Capitalize on the Greatest Business Opportunity of Our Lifetime*, Wiley, Hoboken, New Jersey, 2023.

Ansari, Samaneh, et al., "Feasibility of Keeping Mars Warm with Nanoparticles," in *Science Advances,* vol. 10, no. 32, August 7, 2024, https://www.science.org/doi/10.1126/sciadv.adn4650.

The Antarctic Treaty, 1959, https://documents.ats.aq/recatt/att005_e.pdf.

Arevalo, Evelyn J., "SpaceX Starship Timeline: Making Life Multi-Planetary," *Tesmanian*, November 24, 2019 https://www.tesmanian.com/blogs/tesmanian-blog/spacex-mars-timeline?srsltid=AfmBOoogiwDjLkFwrzRf4Jznk5wbR2NBtx7X6IpR11UyqnY0W8x7tRD4&utm.

Aristo, David, "What's Driving China's Commercial Launch Industry," in SpaceNews, March 2025, 25, https://spacenews.com/whats-driving-chinas-commercial-launch-industry/.

Ausubel, Jesse H., "The Return of Nature. How Technology Liberates the Environment," in *The Breakthrough Journal,* May 12, 2015, https://thebreakthrough.org/journal/issue-5/the-return-of-nature.

Autry, Greg, and Navarro, Peter, *Red Moon Rising: How America Will Beat China on the Final Frontier*, Post Hill Press, New York, Nashville, 2024.

Badescu, Viorel (ed.), *Asteroids: Prospective Energy and Material Resources,* Springer, Heidelberg, New York, Dordrecht, London, 2013.

Barnes, Michael, *SpaceX, Starlink and Our Man Elon Musk: Bringing NASA Astronauts to the Space Station and Mega Project to Colonize Mars*, Oliver Leish, [no place], 2022.

Beech, Martin, *Terraforming: Creating of Habitable Worlds,* Springer, Switzerland, 2009.

Beech, Martin; Seckbach, Joseph; and Gordon, Richard (eds.), *Terraforming Mars*, Wiley, London, 2021.

Bergan, Brad, *Space Race 2.0.: SpaceX, Blue Origin, Virgin Galactic, NASA and the Privatization of the Final Frontier*, Motorbooks, Beverly, MA, 2022.

Berger, Eric, *Reentry: SpaceX, Elon Musk and the Reusable Rockets That Launched a Second Space Age,* BenBella Books, Dallas, TX, 2024.

Berger, Eric, "We Are Going to the Moon," *Reason*, December 2022, https://reason.com/2022/11/15/we-are-going-to-the-moon/.

Billings, Linda, "Should Humans Colonize Mars? No," in *Theology and Science 2019,* vol. 17, no. 3, 341–46, https://doi.org/10.1080/14746700.2019.1632524.

Binder, Jon, *SpaceX Starship: Elon Musk's Rocket to Mars*, Echelon Books, [no place], 2022.

Branson, Richard, *Screw It, Let's Do It: Lessons in Life and Business*, Virgin Books, London, 2009.

Braun, Wernher von, *The Mars Project,* University of Illinois Press, Urbana and Chicago, 1991.

Braun, Wernher von, "Manned Mars Landing, Presentation to the Space Group," August 4, 1969, https://ntrs.nasa.gov/api/citations/20240011937/downloads/1969-08%20MSFC%20von%20Braun%20-%20Manned%20Mars%20Landing.pdf.

Bromberg, Joan Lisa, *NASA and the Space Industry*, John Hopkins University Press, Baltimore and London, 2000.

Byers, Michael, and Boley, Aaaron, *Who Owns Outer Space? International Law, Astrophysics, and the Sustainable Development of Space,* Cambridge University Press, Cambridge u.a., 2023.

Cannon, Kevin M.; Matt, Gialich; and Jose Acain, "Precious and Structural Metals on Asteroids," in *Planetary and Space Science Vol. 225*, 2023 https://www.sciencedirect.com/science/article/pii/S0032063322001945.

Cantrell, Jim, *Breaking All the Rules: The Inside Story of the New Space Race,* [no place], 2023.

Carter, Carl, "History of Space Tourism," in Cohen, Erik, and Spector, Sam (eds.), *Space Tourism: The Elusive Dream*, emerald Publishing United Kingdom, 2019, 61–68.

Chase-Lubitz, Jesse, "Virgin Galactic to Launch Space Tourism Flight as Waiting Lists Grow," *Skift.com*, June 7, 2024,
https://skift.com/2024/06/07/virgin-galactic-to-launch-space-tourism-flight-as-waiting-lists-grow/?utm.

Clary, David A., *Rocket Man: Robert H. Goddard and the Birth of the Space Age*, Grand Central Publishing, New York, 2004.

Cockel, Charles S., "The Ethical Status of Microbial Life on Earth and Elsewhere in Defense of Intrinsic Value," in Schwartz, James S. J., and Milligan, Tony (eds.), *The Ethics of Space Exploration*, Springer International Publishing, Cham, Swizerland, 167–79, https://doi.org/10.1007/978-3-319-39827-3_12.

Cohen, Erik, and Spector, Sam (eds.), *Space Tourism: The Elusive Dream*, emerald Publishing United Kingdom, 2019.

Collins, Patrick, and Autino, Adriano, "What the Growth of a Space Tourism Industry Could Contribute to Employment, Economic Growth, Environmental Protection, Education, Culture and World Peace," in *Acta Astronautica 66*, 2010, 1553–62.

Corrêa, Raphael, *Fundamentals of Space Capitalism: Exploring the Frontiers that Will Shape Tomorrow*, independently published, United States, 2023.

Cucinotta, Francis A., and Durante, Marco, "Cancer Risk from Exposure to Galactic Cosmic Rays: Implications for Space Exploration by Human Beings," *The Lancet Oncology, 7(5)*, 2006, 431–35. https://pubmed.ncbi.nlm.nih.gov/16648048/.

Curcio, Blaine, "China Space in 2024: A Review," in *China Space Monitor*, January 7, 2025, https://chinaspacemonitor.substack.com/p/china-space-in-2024-a-review.

Curcio, Blaine, "Breadth Paired with Depth in China's Space Sector," in *China Space Monitor*, January 8, 2025, https://chinaspacemonitor.substack.com/p/breadth-over-depth-in-chinas-space.

Curcio, Blaine, "The Launch Bottleneck Holding up China's Starlink," in *China Space Monitor*, March 16, 2025, https://chinaspacemonitor.substack.com/p/the-launch-bottleneck-holding-up.

Curcio, Blaine, "51 Rocket Production Sites," in *China Space Monitor*, July 31, 2025, https://chinaspacemonitor.substack.com/p/51-rocket-production-sites.

Daniels, Karen E., "Rubble-Pile Near Earth Objects: Insights from Granular Physics," in Badescu, Viorel (ed.), *Asteroids: Prospective Energy and Material Resources*, Springer, Heidelberg, New York, Dordrecht, London, 2013, 271–86.

Dhir, Rajeev, "Which Country Has the Most Billionaires?," Investopedia, July 2, 2025, https://www.investopedia.com/which-country-has-the-most-billionaires-11752300?utm_.

Dinner, Josh, "Trump Orders Interim NASA Chief to End DEI Initiatives," in SpaceNews, January 23, 2025, https://www.space.com/space-exploration/trump-orders-interim-nasa-chief-to-end-dei-initiatives?utm.

Dripke, Andreas, and Kreft, Prof. Dr. Heinrich, *Race to Space: How Elon Musk, Jeff Bezos, and Richard Branson Are Conquering Space—and the Role of NASA, ESA, China and Russia*, ePub, Diplomatic Council Publishing, Wiesbaden, 2025.

Dubbs, Chris, and Paat-Dahlstrom, Emeline, *Realizing Tomorrow: The Path to Private Spaceflight*, University of Nebraska Press, Lincoln, NE, and London, 2011.

Elvis, Martin, *Asteroids: How Love, Fear, and Greed Will Determine Our Future in Space*, Yale University Press, New Haven & London 2021.

European Space Agency (ESA), "Diversity and Inclusiveness," [no date], https://www.esa.int/About_Us/Careers_at_ESA/Diversity_and_Inclusiveness?utm.

Fehrler, Sebastian; Hornuf, Lars; and Vrankar, Daniel, "What Do Citizens Expect from Space?" *Acta Astronautica*, vol. 235, October 2025, 660–69, https://doi.org/10.1016/j.actaastro.2025.05.056

Fleming, Maxwell; Lange, Ian; Shojaeinia, Sayeh; and Stuermer, Martin, "Mining in Space Could Spur Sustainable Growth," in *PNAS 2023,* vol. 120, no. 43, https://www.pnas.org/doi/10.1073/pnas.2221345120.

Fogg, Martyn J., "The Ethical Dimensions of Space Settlement," in *Space Policy 16,* 2000, 205–11, https://doi.org/10.1016/S0265-9646(00)00024-2.

Follett, Andrew, "7 Enviro Predictions From Earth Day 1970 That Were Just Dead Wrong," *Daily Caller*, April 22, 2016, https://dailycaller.com/2016/04/22/7-enviro-predictions-from-earth-day-1970-that-were-just-dead-wrong/.

Foust, Jeff, "Debating the Aldridge report," *The Space Review*, August 30, 2005, https://www.thespacereview.com/article/217/1?utm_.

Foust, Jeff, "The Accidental Monopoly. How SpaceX Became (just about) the Only Game in Town," SpaceNews, October 13, 2023, https://spacenews.com/the-accidental-monopoly/.

Foust, Jeff, "NASA's Budget Crisis Presents an Opportunity for Change," SpaceNews, June 4, 2025, https://spacenews.com/nasas-budget-crisis-presents-an-opportunity-for-change/.

Genta, Giancarlo, "Private Space Exploration: A New Way for Starting a Spacefaring Society?" *Acta* Astronautica, vol. 104, no. 2, 2014, 480–86, https://doi.org/10.1016/j.actaastro.2014.04.008.

Genta, Giancarlo, "Terraforming and Colonizing Mars," in Beech, Martin; Seckbach, Joseph; and Gordon, Richard (eds.), *Terraforming Mars*, Wiley, London, 2021, 3–22.

Gouw, Arwin M., "CRISPR Challenges and Opportunities for Space Travel," in Konrad Szocik (ed.), *Human Enhancement for Space Missions: Lunar, Martian, and Future Missions to the Outer Planets,* Springer, Cham, Switzerland, 2020.

Grandl, Werner, and Bazso, Akos, "Near Earth Asteroids—Prospection, Orbit Modification and Habitation," in Badescu, Viorel (ed.), *Asteroids. Prospective Energy and Material Resources,* Springer, Heidelberg, New York, Dordrecht, London, 2013, 415–38.

Green, Brian Patrick, *Space Ethics,* Bloomsbury Publishing, London, 2021.

Gregg, Jack, *The Cosmos Economy: The Industrialization of Space*, Springer Nature, Cham, Switzerland, 2021.

Guthrie, Julian, *How to Make a Spaceship*, Penguin Press, New York, 2016.

Hajduk, Chris, "The First Settlement of Mars," in Beech, Martin; Seckbach, Joseph; and Gordon, Richard (eds.), *Terraforming Mars*, Wiley, London, 2021, 331–51.

Hallion, Richard P., *Taking Flight: Inventing the Aerial Age from Antiquity through the First World War*, Oxford University Press, New York, 2003.

Hamill, Doris; Mongan, Philipp; and Kearney, Michael, "Space Commerce: An Entrepreneur's Angle," in Hudgins, Edward L. (ed.), *Space: The Free Market Frontier,* CATO Institute, Washington, 2002, 151–66.

Handberg, Roger, *The Future of the Space Industry: Private Enterprise and Public Policy,* Quorum Books, Westport, Connecticut, London, 1995.

Hanley, Steve, "Reusable Rockets Are All the Rage in China. Now Honda Wants in on the Fun," *CleanTechnica*, June 22, 2025, https://cleantechnica.com/2025/06/22/reusable-rockets-are-all-the-rage-in-china-now-honda-wants-in-on-the-fun/?utm.

Harvey, Brian, *China in Space: The Great Leap Forward. Second Edition*, Springer Praxis Publishing, Chichester, UK, 2019.

Heinlein, Robert A., *The Man Who Sold the Moon*, New American Library, New Jersey, 1951.

Hennigan, W.J., "Astronauts Take Shelter Aboard ISS after Russian Anti-Satellite Test, US Says," in *Time*, November 15, 2021, https://time.com/6117840/astronauts-shelter-iss-russia-test/?utm.

Henrekson, Magnus; Sandström, Christian; and Mikael Stenkula (eds.), *Moonshots and the New Industrial Policy: Questioning the Mission Economy*, Springer Nature, Cham, Switzerland, 2024.

Henrekson, Magnus; Sandström, Christian; and Mikael Stenkula, "Learning from Overrated Mission-Oriented Innovation Policies: Seven Takeaways," in Henrekson, Magnus; Sandström, Christian; and Mikael Stenkula (eds.), *Moonshots and the New Industrial Policy: Questioning the Mission Economy*, Springer Nature, Cham, Switzerland, 2024, 234–55.

Hersch, Matthew H., *Dark Star: A New History of the Space Shuttle*, The MIT Press, Cambridge, MA, London, 2023.

Hornsey, Matthew; Fielding, Kelly S.; Harris, Emily A.; Bain, Paul G.; Grice, Tim; and Chapman, Cassandra M., "Protecting the Planet or Destroying the Universe? Understanding Reactions to Space Mining," in *Sustainability* vol. 14, no. 7, March 30, 2022, https://doi.org/10.3390/su14074119.

Hornuf, Lars, and Vrankar, Daniel, "How New Business Models Shape Innovation Spillovers: Insights from the New Space Economy," *SSRN*, June 14, 2025, https://papers.ssrn.com/sol3/papers.cfm?abstract_id=5294579&utm.

Howell, Elizabeth, "SpaceX's Dragon: First Private Spacecraft to Reach the Space Station," Space.com, August 10, 2020, https://www.space.com/18852-spacex-dragon.html.

Hudgins, Edward L. (ed.), *Space: The Free Market Frontier*, CATO Institute, Washington, 2002.

IATA, "IATA Releases 2024 Safety Report," February 26, 2025, https://www.iata.org/en/pressroom/2025-releases/2025-02-26-01/?utm.

Idealista News, "Moon States oder wie man Grundstücke auf dem Mond für 34 € verkauft," Idealista News, August 31, 2023,https://www.idealista.com/de/news/immobilien-kaufen-in-spanien/2023/08/31/140361-moon-states-oder-wie-man-grundstuecke-auf-dem-mond-fuer-34-eu-verkauft.

Isaacson, Walter, *Elon Musk*, Simon & Schuster, New York, 2023.

Jones, Andrew, "China's Megaconstellation Launches Could Litter Orbit for More than a Century, Analysts Warn," SpaceNews, April 7, 2025,https://spacenews.com/chinas-megaconstellation-launches-could-litter-orbit-for-more-than-a-century-analysts-warn/.

Jones, Harry W. (NASA Ames Research Center), *The Recent Large Reduction in Space Launch Cost*, 48th International Conference on Environmental Systems, 8–12 July 2018, Albuquerque, New Mexico, https://ntrs.nasa.gov/api/citations/20200001093/downloads/20200001093.pdf.

Jones, Harry W., "Current Capabilities Can Get Us to Mars," in *AIAA ARC Aerospace Research Central*, July 27, 2024, https://arc.aiaa.org/doi/abs/10.2514/6.2024-4862.

Jones, Harry W., (NASA Ames Research Center), "The Impact of Reduced Space Launch Cost," *AIAA ARC Aerospace Research Central*, July 16 2025, https://arc.aiaa.org/doi/10.2514/6.2025-4073.

Kaku, Michio, *The Future of Humanity: Our Destiny in the Universe,* Vintage Books, New York, 2019.

Kang, Si-Yoon; Jo, Min-Seon; Choi, Jeong-Yeol; and Yang, Soo Seok, "Cost Effectiveness of Reusable Launch Vehicles Depending on the Payload Capacity," *Aerospace,* vol. 12, no. 5, April 22, 2025, 364, https://doi.org/10.3390/aerospace12050364.

Kluger, Jeffrey, "Space: Where America and Russia Are Stuck with Each Other," *Time*, March 25, 2014, https://time.com/37671/space-cooperation-america-russia/.

Kluger, Jeffrey, "Russia, We Have a Problem," *Time*, June 12, 2014, https://time.com/2863223/russia-we-have-a-problem/.

Koelle, Dietrich E., and Janovsky, R., "Development and Transportation costs of Space Launch Systems, presentation at the 1st CEAS European Air and Space Conference, Berlin, 2007, https://www.fzt.haw-hamburg.de/pers/Scholz/ewade/2007/CEAS2007/papers2007/ceas-2007-289.pdf.

Kreutzer, Ralf T., and Land, Karl-Heinz, *Dematerialisierung: Die Neuverteilung der Welt in Zeiten des digitalen Darwinismus*, Future Vision Press, Cologne, 2015.

Kuczera, Heribert, and Sacher, Peter W., *Reusable Space Transportation Systems,* Springer Praxis Publishing, Chichester, UK, Heidelberg, Germany, 2011.

Laing, Jennifer, and Frost, Warwick, "Exploring Motivations of Potential Space Tourists," in Cohen, Erik, and Spector, Sam (eds.), *Space Tourism: The Elusive Dream,* emerald Publishing United Kingdom, 2019, 141–62.

Letizia, Anthony, "Congressman James Fulton and U.S. Space Policy," *Geek Frontiers*, July 14, 2019, https://geekfrontiers.com/congressman-james-fulton-and-us-space-policy/.

Lewis, John S., *Mining the Sky: Untold Riches from the Asteroids, Comets and Planets*, Helix Books Basic, New York, 1997.

Lewis, John S., *Asteroid Mining 101: Wealth for the New Space Economy*, Deep Space Industries, Moffett Field, CA, 2015.

Librera, Gianluca, "Aerospace Technologies and Civil Spillovers," Master's thesis, Politecnico di Torino, 2022, https://webthesis.biblio.polito.it/25081/.

Logsdon, John M., *After Apollo? Richard Nixon and the American Space Program,* Palgrave Macmillan, New York, 2015.

Lorenzen, Dirk, H., *Die Zukunft der Raumfahrt: Der neue Wettlauf ins All*, Franck Kosmos Verlag, Stuttgart, 2021.

Lowe, Rebecca, *Space Invaders: Property Rights on the Moon*, Adam Smith Research Trust, London, 2022.

Luscombe, Richard, "'We Weren't Stuck': Nasa Astronauts Tell of Space Odyssey and Reject Claims of Neglect," *The Guardian*, March 31, 2025, https://www.theguardian.com/science/2025/mar/31/nasa-astronauts-iss-trump-musk?utm_.

MacDonald, Alexander, *The Long Space Age: The Economic Origins of Space Exploration from Colonial America to the Cold War*, Yale University Press, New Haven, CT, and London, 2017.

Mangu-Ward, Katherine, "From Space Regulator to Astronaut: Q & A Interview with George Nield," *Reason*, December 2022, https://reason.com/2022/11/15/from-space-regulator-to-astronaut/.

Mangu-Ward, Katherine, "The Case for Space Billionaires," *Reason*, December 2022, https://reason.com/2022/11/15/the-case-for-space-billionaires/.

Mazzucato, Mariana, *The Value of Everything: Making and Taking in the Global Economy*, PublicAffairs, New York, 2018.

Mazzucato, Mariana, *Mission Economy: A Moonshot Guide to Changing Capitalism,* Penguin Books, London, 2022.

McAfee, Andrew, *The Surprising Story of How We Learned to Prosper Using Fewer Resources—and What Happens Next*, Scribner, New York, 2019.

McCloskey, Deirdre Nansen, and Mingardi, Alberto, *The Myth of the Entrepreneurial State*, American Institute for Economic Research, Great Barrington, MA, Adam Smith Institute, London, 2020.

McCurdy, Howard E., *The Space Station Decision: Incremental Politics and Technological Choice*, Johns Hopkins University Press, Baltimore, 1990.

McDougall, Walter A., *The Heavens and the Earth: A Political History of the Space Age*, Johns Hopkins University Press, Baltimore, 1985.

McKaig, Jordan; Caro, Tristan; Burton, Dana; Tavares, Frank; and Vidaurri, Monica, "Chapter 10: Planetary Protection—History, Science, and the Future," *Astrobiology,* vol. 24, no. S1, March 2024, https://doi.org/10.1089/ast.2021.0112.

McKenzie, Richard B., and Lee, Dwight R., *In Defense of Monopoly: How Market Power Fosters Creative Production*, University of Michigan Press, Ann Arbor, MI, 2008.

Merchant, Brian, "The Problem with Asteroid Mining," Vice.com, January 23, 2013, https://www.vice.com/en/article/the-problem-with-asteroid-mining/.

Milligan, Tony, *Nobody Owns the Moon: The Ethics of Space Exploitation*, McFarland & Company, Jefferson, NC, 2015.

Mingardi, Alberto, "A Critique of Mazzucato's Entrepreneurial State," *Cato Journal,* vol. 35, no. 3, Fall 2015, https://www.cato.org/sites/cato.org/files/serials/files/cato-journal/2015/9/cj-v35n3-7.pdf.

Miraux, Lois; Wilson, Andrew Ross; and Dominguez Calabuig, Guillermo J., "Environmental Sustainability of Future Proposed Space Activities," *Acta Astronautica,* vol. 200, November 2022, 329–46, https://doi.org/10.1016/j.actaastro.2022.07.034.

Morino, Lori, "Humanity Is Not Prepared to Colonize Mars," *Futures*, vol. 110, June 2019, 15–18, https://doi.org/10.1016/j.futures.2019.02.010.

Munévar, Gonzalo, "An Obligation to Colonize Outer Space," *Futures,* vol. 110, June 2019, 38–40, https://doi.org/10.1016/j.futures.2019.02.009.

Munévar, Gonzalo, *The Dimming of Starlight: The Philosophy of Space Exploration*, Oxford University Press, Oxford, UK, 2023.

Musk, Elon, "Full interview with Kara Swisher, and Walt Mossberg," Code Conference 2016, June 2016, https://www.youtube.com/watch?v=wsixsRI-Sz4.

Musselman, Brain T.; Winter, Scott R.; Rice, Stephen; Keebler, Joseph R.; and Ruskin, Keith J., "Point-to-Point Suborbital Space Tourism Motivation and Willingness to Fly," *Annals of Tourism Research Empirical Insights,* vol. 5, no. 1, May 2024,https://doi.org/10.1016/j.annale.2024.10011.

Mwai, Peter; Gebru, Girmay; and Thomas, Merlyn, "Satellite Images and Doctor Testimony Reveal Tigray Hunger Crisis," BBC, July 25, 2024, https://www.bbc.com/news/articles/c10l2vvjy9lo.

NASA, "NASA's Efforts to Maximize Research on the International Space Station, Audit Report, July 8, 2013, Report No. IG-13-019, https://oig.nasa.gov/office-of-inspector-general-oig/ig-13-019/.

NASA, "The Artemis Accords: Principles for Cooperation in the Civil Exploration and Use of the Moon, Mars, Comets and Asteroids for Peaceful Purposes, October 13, 2020, https://www.nasa.gov/wp-content/uploads/2022/11/Artemis-Accords-signed-13Oct2020.pdf?utm.

NASA, "Psyche Mission Overview," nasa.gov, April 17, 2025, https://science.nasa.gov/mission/psyche/mission-overview/#:~:text=Psyche%20is%20a%20NASA%20mission,metal%20than%20rock%20or%20ice.

National Academics of Sciences, Engineering and Medicine, "Space Radiation and Astronaut Health: Managing and Communicating Cancer Risks," The National Academies Press, Washington DC, 2021,https://doi.org/10.17226/26155.

Nelson, Peter Lothian, and Block, Walter E., *Space Capitalism: How Humans Will Colonize Planets, Moons, and Asteroids,* Palgrave Macmillan, Cham, Switzerland, 2018.

Nichols-Fleming, Fiona; Evans, Alexander J.; Johnson, Brandon C.; and Sori, Michael M., "Porosity Evolution in Metallic Asteroids: Implications for the Origin and Thermal History of Asteroid 16 Psyche," *JGR Planets,* vol. 127, no. 2, February 2022, https://doi.org/10.1029/2021JE007063.

Nötzold, Antje; Fels, Enrico; Rotter, Andrea; and Brake, Moritz, *Strategischer Wettbewerb im Weltraum: Politik, Recht, Sicherheit und Wirtschaft im All,* Springer VS, Wiesbaden, 2024.

Ormrod, James, "Pro-space Activism and Narcissistic Phantasy," *Psychoanalysis, Culture and Society,* vol. 12, September 1, 2007, 260–78, https://doi.org/10.1057/palgrave.pcs.2100131.

Ormrod, James; Dickens, Peter, "Space Tourism, Capital, and Identity," Cohen, Erik, and Spector, Sam (eds.), *Space Tourism: The Elusive Dream*, emerald Publishing United Kingdom, 2019, 223–44.

Orosai, Roberto, et al., "Radar Evidence of Subglacial Liquid Water on Mars," *Science,* vol. 381, no. 6401, July 25, 2018, 490–93, https://www.science.org/doi/10.1126/science.aar7268.

Partouche, Alexa, "SpaceX Registers to Build 700,000-square-foot Starship-Producing 'Gigabay' in Starbase," *Houston Chronicle,* July 9, 2025, https://www.houstonchronicle.com/news/houston-texas/trending/article/spacex-starbase-gigabay-facility-20763151.php?utm.

Peters, Björn, *Schluss mit der Energiewende! Warum Deutschlands Volkswirtschaft dringend ökologischen Realismus braucht*, Orgshop Verlag, Moos, Germany, 2025.

Pippo, Simonetta Di, *Space Economy: The New Frontier for Development*, Bocconi University Press, Milan, 2023.

The Planetary Society, "How Much Did the Apollo Program Cost?" planetary.org, [no date], https://www.planetary.org/space-policy/cost-of-apollo?utm.

Poole, Robert, "Robert Heinlein's Dream of Private Space Travel is Coming True," *Reason*, vol. 54, no. 7, December 2022, https://reason.com/2022/11/15/we-are-living-robert-heinleins-dream/.

Poole, Robert, "Is This Any Way to Run Space Transportation?" in Hudgins, Edward L. (ed.), Space: The *Free Market Frontier,* CATO Institute, Washington, 2002, 53–66.

Pop, Virgiliu, *Unreal Estate: The Man Who Sold the Moon*, Lulu.com, Morrisville, NC, 2006.

Pop, Virgiliu, "Legal Considerations on Asteroid Exploitation and Deflection," in Badescu, Viorel (ed.), *Asteroids: Prospective Energy and Material Resources*, Springer, Heidelberg, New York, Dordrecht, London, 2013, 659–80.

Pronczuk, Monika, "Europe Wants to Diversify Its Pool of Astronauts," *The New York Times*, February 22, 2021, https://www.nytimes.com/2021/02/22/world/europe/women-disabled-astronauts.html.

The Protocol on Environmental Protection to the Antarctic Treaty, 1991, https://documents.ats.aq/recatt/Att006_e.pdf.

Pultarova, Tereza, "China's Push for a More Commercial Space Industry," *Via Satellite*, May 28, 2024, https://interactive.satellitetoday.com/via/june-2024/chinas-push-for-a-more-commercial-space-industry.

Rainbow, Jason, "The Efforts Bridging Space Sustainability, from Best Intentions to Real-World Actions," SpaceNews, February 10, 2025, https://spacenews.com/the-efforts-bridging-space-sustainability-from-best-intentions-to-real-world-actions/.

Rainbow, Jason, "SN Intelligence from SpaceNews: Understanding the SpaceX-Era Economy, Part 1: Launch Supremacy," SpaceNews, July 21, 2025.

Reagan, Ronald, "Radio Address to the Nation on the Space Program," January 28, 1984, National Archives, Ronald Reagan Library & Museum, https://www.reaganlibrary.gov/archives/speech/radio-address-nation-space-program?utm.

Reichl, Eugen, *Chinas Raumfahrt: Ein Riese erwacht*, Motorbuch Verlag, Stuttgart, 2023.

Reichl, Eugen, *Die Zukunft der Raumfahrt: Private Projekte*, Motorbuch Verlag, Stuttgart, 2022.

Reichl, Eugen, *Menschen im Weltraum*, Springer, Berlin, 2022.

Reichl, Eugen, *Space 2022: Das aktuelle Raumfahrt-Jahr mit Chronik 2021,* VfR, Munich, 2021.

Reichl, Eugen, *Space 2024: Das aktuelle Raumfahrt-Jahrbuch mit allen Starts*, VfR, Munich, 2023.

Reichl, Eugen, *Space 2025: Das aktuelle Raumfahrt-Jahrbuch mit allen Starts*, VfR, Munich, 2024.

Reichl, Eugen, *Raumfahrt-Geschichte: Die 100 wichtigsten Ereignisse*, Motorbuch Verlag, Stuttgart, 2021.

Reinstein, Ezra J., "Owning Outer Space," *Northwestern Journal of International Law & Business*, vol. 20, no. 1, Fall 1999, 59–90, https://scholarlycommons.law.northwestern.edu/njilb/vol20/iss1/7/.

Reynolds, Glenn H., and Merges, Robert P., *Outer Space: Problems of Law and Policy, 2nd Edition*, Routledge, London, New York, 2019.

Roettgen, Raphael, *To Infinity: The Space Economy & How You Can Participate*, Space Business Institute, [no place], 2024.

Rubenstein, Mary-Jane, *Astrotopia: The Dangerous Religion of the Corporate Space Race*, The University of Chicago Press, Chicago, London, 2022.

Ryan, Mike H., and Kutschera, Ida, "The Case of Asteroids," in Badescu, Viorel (ed.), *Asteroids: Prospective Energy and Material Resources*, Springer, Heidelberg, New York, Dordrecht, London, 2013, 645–58.

Salvanto, Anthony, "CBS News Poll: Most Americans Favor U.S. Returning to Moon, going to Mars," CBS News, July 18, 2025, [online]https://www.cbsnews.com/news/cbs-news-poll-most-americans-favor-u-s-returning-to-moon-going-to-mars/.

Schladebach, Marcus, *Weltraumrecht*, Mohr Siebeck, Tübingen, Germany, 2020.

Schlather, Marc, "The Legislative Challenge in Space Transportation Financing," in Hudgins, Edward L. (ed.), *Space: The Free Market Frontier*, CATO Institute, Washington, 2002, 195–212.

Schmidt, Samantha, "The Gold-Mining City that Is Destroying a Sacred Venezuelan Mountain," *The Washington Post*, December 6, 2022, https://www.washingtonpost.com/world/2022/12/06/venezuela-yapacana-gold-mining/.

Schneider, Peter M., *Goldrausch im All: Wie Elon Musk, Richard Branson und Jeff Bezos den Weltraum erobern*, FinanzBuch Verlag, Munich, 2018.

Schwartz, James S. J., *The Value of Science in Space Exploration*, Oxford University Press, Oxford, UK, 2020.

Schwartz, James S. J., and Milligan, Tony, *The Ethics of Space Exploration*, Springer, Cham, Switzerland, 2016.

Scotti, Monique, "NASA Plans Mission to a Metal-Rich Asteroid Worth Quadrillions," *Global News*, January 14, 2017, https://globalnews.ca/news/3175097/nasa-plans-mission-to-a-metal-rich-asteroid-worth-quadrillions/.

Seedhouse, Erik, *SpaceX: Starship to Mars – The First 20 Years, 2nd Edition*, Springer, Cham, Switzerland, 2022.

Simberg, Rand, "Homesteading the Final Frontier. A Practical Proposal for Securing Property Rights in Space," *CEI Competitive Enterprise Institute, Issue Analysis, April 2012, No. 3*, https://cei.org/sites/default/files/Rand%20Simberg%20-%20Homesteading%20the%20Final%20Frontier.pdf.

Simberg, Rand, *Safe Is Not an Option: Overcoming the Futile Obsession with Getting Everyone Back Alive That Is Killing Our Expansion into Space*, Interglobal Media LLC, Jackson, WY, 2014.

Singh, Shalini, and Misra, Pankaj, "Mapping the Thematic Evaluation and Future Directions of Space Tourism Research," *New Space*, June 13, 2025, https://www.researchgate.net/publication/392679376_Mapping_the_Thematic_Evaluation_and_Future_Directions_of_Space_Tourism_Research.

Sivolella, Davide, *Space Mining and Manufacturing: Off-World Resources and Revolutionary Engineering Techniques*, Springer Nature, Cham, Switzerland, 2019.

Smiles, Deondre, "The Settler Logics of (Outer) Space," *Society and Space*, October 26, 2020, https://www.societyandspace.org/articles/the-settler-logics-of-outer-space.

Smith, Kelley C., et al., "The Great Colonization Debate," *Futures,* vol. 110, June 2019, https://doi.org/10.1016/j.futures.2019.02.004, 4–14.

Smith, Kiona N., "The Correction Heard 'Round The World: When the New York Times Apologized to Robert Goddard," *Forbes*, July 19, 2018, https://www.forbes.com/sites/kionasmith/2018/07/19/the-correction-heard-round-the-world-when-the-new-york-times-apologized-to-robert-goddard/.

Smith, Marcia, "NASA, ROSCOSMOS Agree on One More Soyuz Seat," *Space Policy Online*, May 12, 2020, https://spacepolicyonline.com/news/nasa-roscosmos-agree-on-one-more-soyuz-seat/.

Space Ambition, and Kalinin, Denis, "China: Private Space Ecosystem of the Rising Superpower," *Space Ambition*, April 25, 2025, https://spaceambition.substack.com/p/china-private-space-ecosystem-of.

Space Capital, "Space Investment Quarterly: Q2 2025," July 10, 2025, https://www.spacecapital.com/publications/space-investment-quarterly-q2-2025.

Sparrow, Robert, "The Ethics of Terraforming," *Environmental Ethics,* vol. 21, no. 3, Fall 1999, 227–45, https://doi.org/10.5840/enviroethics199921315.

Spector, Sam, and Higham, James E. S., "Space Tourism, the Anthropocene, and Sustainability," in Cohen, Erik, and Spector, Sam (eds.), *Space Tourism: The Elusive Dream*, emerald Publishing United Kingdom, 2019, 245–62.

Stoner, Ian, "The Ethics of Terraforming: A Critical Survey of Six Arguments," in Beech, Martin; Seckbach, Joseph, and Gordon, Richard (eds.), *Terraforming Mars*, Wiley, London, 2021, 101–16.

Storck, Lisa Maria, "Space Mining—Goldrausch im All? Eine meta-geopolitische Analyse der strategischen Bedeutung von Weltraumbergbau für die internationale Staatengemeinschaft," in Nötzold, Antje; Fels, Enrico; Rotter, Andrea, and Brake, Moritz (eds.), *Strategischer Wettbewerb im Weltraum: Politik, Recht, Sicherheit und Wirtschaft im All*, Springer VS, Wiesbaden, Germany, 2024, 367–87.

Stuhlinger, Ernst, and Ordway, Frederick I., *Wernher von Braun: Crusader for Space: A Biographical Memoir,* Malabar, Florida, 1994.

Synergy, and Guest Author, "The Great Leap Forward of China's Private Space Industry," *Synergy, The Journal of Contemporary Asian Studies*, April 9, 2025, https://utsynergyjournal.org/2025/04/09/the-great-leap-forward-of-chinas-private-space-industry/.

Szocik, Konrad (ed.), *Human Enhancement for Space Missions: Lunar, Martian, and Future Missions to the Outer Planets*, Springer Nature, Cham, Switzerland, 2020.

Tavares, Frank, et al., "Ethical Exploration and the Role of Planetary Protection in Disrupting Colonial Practices. A Submission to the Planetary Science and Astrobiology Decadal Survey 2023–2032," *arXiv*, October 27, 2020, https://doi.org/10.48550/arXiv.2010.08344.

Thomas, Gareth Morgan, *The Chinese Space Dynasty: China's Robotic Revolution and the New Space Age,* Burst Book, 2025.

Toivonen, Annette, *Sustainable Space Tourism: An Introduction*, Channel View Publications, Bristol, UK, 2020.

United Nations Office for Outer Space Affairs (UNOOSA), "Moon Agreement," https://www.unoosa.org/oosa/en/ourwork/spacelaw/treaties/moon-agreement.html.

United Nations Office for Outer Space Affairs (UNOOSA), "Treaty on Principles Governing the Activities of States in the Exploration and Use of Outer Space, including the Moon and other Celestial Bodies," https://www.unoosa.org/pdf/gares/ARES_21_2222E.pdf.

Unnerstall, Thomas, "Liebe Schüler, ihr verbraucht drei Erden!," in Benkens, Robert; Bojanowski, Axel; and von Storch, Hans (eds.), *Das andere Klimabuch: 25 Experten, überraschende Fakten, rationale Alternativen zur Klimaangst*, Königshausen & Neumann Verlag, Würzburg, Germany, 2025, 65–75.

UNOOSA, United Nations Office for Outer Space Affairs, "2222 (XXI) Treaty on Principles Governing the Activities of States in the Exploration and Use of Outer Space, including the Moon and other Celestial Bodies, December 19, 1966, https://www.unoosa.org/oosa/en/ourwork/spacelaw/treaties/outerspacetreaty.html.

Vance, Ashlee, *Elon Musk: How the Billionaire CEO of SpaceX and Tesla Is Shaping Our Future*, Virgin Books, London 2016.

Vance, Ashlee, *When the Heavens Went on Sale: The Misfits and Geniuses Racing to Put Space Within Reach*, Ecco, New York, 2023.

Vidaurri, Monica, et al., "Absolute Prioritization of Planetary Protection, Safety, and Avoiding Imperialism in all Future Science Missions: A Policy Perspective," *Space Policy,* vol. 51, no. 3, March 2020, 1–6, https://doi.org/10.1016/j.spacepol.2019.101345.

Viiera Neto, Ernesto; Pires, Pryscilla; and Giuliatti Winter, Silvia, "Trajectories for Mining Space Mission Asteroids in Near-Earth Orbit," *The European Physical Journal Special Topics,* vol. 232, December 7, 2023, 2967–74, https://link.springer.com/article/10.1140/epjs/s11734-023-01016-y.

Wall, Mike, "NASA's Shuttle Program Cost $209 Billion—Was It Worth it?" Space.com, July 5, 2011, https://www.space.com/12166-space-shuttle-program-cost-promises-209-billion.html?utm.

Wanjek, Christopher, *Spacefarers: How Humans Will Settle the Moon, Mars, and Beyond*, Harvard University Press, Cambridge, MA, London, 2020.

Wasser, Alan, and Jobes, Douglas L., "Space Settlements, Property Rights, and International Law: Could a Lunar Settlement Claim the Lunar Real Estate it Needs to Survive?" *Journal of Air Law and Commerce,* vol. 73, no. 1, 2008, https://www.space-settlement-institute.org/Articles/jal73-1Wasser.pdf.

Webber, Derek, "Current Space Tourism Developments," in Cohen, Erik, and Spector, Sam (eds.), *Space Tourism: The Elusive Dream*, emerald Publishing United Kingdom, 2019, 163–76.

Weinersmith, Kelly, Weinersmith, Zach, *A City on Mars: Can We Settle Space, Should We Settle Space, and Have We Really Thought This Through?*, Penguin, London, 2024.

Weinzierl, Matthew, and Rosseau, Brendan, *Space to Grow: Unlocking the Final Economic Frontier*, Harvard Business Review Press, Boston, MA, 2025.

Wemheuer, Felix, *Der große Hunger: Hungersnöte unter Stalin und Mao*, Rotbuch Verlag, Berlin, 2012.

Wikipedia, "Outer Space Treaty," wikipedia.org, [no date], https://en.wikipedia.org/wiki/Outer_Space_Treaty?utm_.

Williams, Walter E., "Environmentalists Are Dead Wrong," creators.com, April 26, 2017, https://www.creators.com/read/walter-williams/04/17/environmentalists-are-dead-wrong.

Worden, S.P., "On Self-Licking Ice Cream Cones," *ASP Conference Series, Vol. 26*, 1992, 600–601, https://adsabs.harvard.edu/full/1992ASPC.26.599W.

World Bank and Development Research Center of the State Council, the People's Republic of China, "Four Decades of Poverty Reduction in China: Drivers, Insights for the World, and the Way Ahead," Washington DC, 2022, https://documents1.worldbank.org/curated/en/099540207212222589/pdf/IDU0d81807d20c45a041160863409697a9fc59c4.pdf.

World Economic Forum, "How China's New Generation of Companies Can Embrace Global Opportunities," June 24, 2025, https://www.weforum.org/stories/2025/06/how-chinas-new-generation-of-companies-can-embrace-global-opportunities/?utm_.

Wright, Helen, *James Lick's Monument: The Saga of Captain Richard Floyd and the Building of the Lick Observatory*, Cambridge University Press, Cambridge, London, New York, New Rochelle, Melbourne, Sydney, 1987.

Young, James Harvey, *The Toadstool Millionaires. A Social History of Patent Medicines in America Before Federal Regulation*, Princeton University Press Princeton, NJ, 1979.

Zhang, Weiying, "The China Model View Is Factually False," *Journal of Chinese Economic and Business Studies,* vol. 17, no. 3: *China's Path to the New Era*, 2019, https://www.tandfonline.com/doi/full/10.1080/14765284.2019.1663696?scroll=top&needAccess=true.

Zhang, Weiying, *The Logic of the Market: An Insider's View of Chinese Economic Reform*, Cato, Washington, DC, 2015.

Zhang, Weiying, *Re-Understanding Entrepreneurship: What It Is and Why It Matters*, Cambridge University Press, Cambridge, UK, New York, 2024.

Zitelmann, Rainer, *The Power of Capitalism: Capitalism is Not the Problem, But the Solution*, Management Books 2000, London, 2024.

Zitelmann, Rainer, "Anti-Capitalists, Post-Colonialists, and the Controversy about the 'Colonisation of Space,'" *Economic Affairs,* vol. 44, no. 3., October 3, 2024, 572–81, https://onlinelibrary.wiley.com/doi/10.1111/ecaf.12672.

Zitelmann, Rainer, *In Defense of Capitalism: Debunking the Myths*, Republic Book Publishers, New York, 2023.

Zitelmann, Rainer, "Be Careful Trump. Deporting Elon Musk Would Hand Space Travel to China," *City AM*, July 2, 2025, https://www.cityam.com/be-careful-trump-deporting-elon-musk-would-hand-space-travel-to-china/.

Zitelmann, Rainer, *How Nations Escape Poverty: Vietnam, Poland and the Origins of Prosperity*, Encounter Books, New York, London, 2024.

Zitelmann, Rainer, *2075: When Beauty Became a Crime*, MB2000, Oxford, 2026.

Zubrin, Robert (with Richard Wagner), *The Case for Mars: The Plan to Settle the Red Planet and Why We Must*, 25th Edition, Free Press, New York, and London, 2021.

Zubrin, Robert, *The Case for Space: How the Revolution in Spaceflight Opens up a Future of Limitless Possibility*, Prometheus Books, Guilford, CT, 2019.

Zubrin, Robert, "A Declaration of Decadence," in *Quillette*, May 30, 2023, https://quillette.com/2023/05/30/a-declaration-of-decadence/.

Zubrin, Robert, "Why We Should Settle Mars," in *Quillette*, December 4, 2023, https://quillette.com/2023/12/04/why-we-should-go-to-mars/.

Zubrin, Robert. *The New World on Mars: What We Can Create on the Red Planet*, Diversion Books, New York, 2024.

ABOUT THE AUTHOR

Rainer Zitelmann is a historian and sociologist specializing in economics and economic history. Born in Frankfurt am Main in 1957, he studied history and political science in Darmstadt from 1978 to 1983, graduating with honors. In 1986, he earned his doctorate under Prof. Dr. Dr. h.c. K. O. Frhr. von Aretin, with his thesis on Hitler's social and economic worldview (*Hitler's National Socialism*). This thesis, which was awarded summa cum laude, garnered international acclaim and was translated into multiple languages.

From 1987 to 1992, Zitelmann worked at the Central Institute for Social Science Research at the Free University of Berlin. He then took on the role of Chief Editor at Ullstein-Propyläen-Verlag, then Germany's third-largest publishing group, and headed various departments at the daily newspaper *Die Welt* until 2000. In 2000, he founded his own business, Dr. ZitelmannPB. GmbH, which has since become the market leader in positioning consulting for real estate companies in Germany. In 2016, he sold the company.

In 2016, he earned a second doctorate, this time in sociology, from the University of Potsdam. This second dissertation, published under the title *The Wealth Elite*, has been published in Germany, the United States, China, Russia, South Korea, and Vietnam.

To date, Zitelmann has authored and edited thirty-two books, which have been translated into thirty-five languages. He is a highly sought-after speaker across Asia, the United States, Latin America, and Europe. In recent years, his articles and interviews have appeared in prestigious media outlets such as the *Wall Street Journal*, *Newsweek*, *Forbes*, *The Daily Telegraph*, *The Times*, *Le Monde*, *Corriere de la Serra*, *Frankfurter Allgemeine Zeitung*, *Der Spiegel*, *Neue Zürcher Zeitung*, and numerous media outlets in Asia and Latin America. He regularly contributes specialist articles to *Economic Affairs* and has coproduced films on economic development in Germany, Poland, Vietnam, and Sweden.

INDEX OF PERSONS